CONTROVERSIAL ENCOUNTERS IN THE AGE OF ALGORITHMS

How Digital Technologies Are Stifling Public Debate and What to Do About It

Sine N. Just

First published in Great Britain in 2025 by

Bristol University Press
University of Bristol
1-9 Old Park Hill
Bristol
BS2 8BB
UK
t: +44 (0)117 374 6645
e: bup-info@bristol.ac.uk

Details of international sales and distribution partners are available at bristoluniversitypress.co.uk

British Library Cataloguing in Publication Data
A catalogue record for this book is available from the British Library

ISBN 978-1-5292-3834-1 paperback
ISBN 978-1-5292-3835-8 OA Pub
ISBN 978-1-5292-3836-5 OA PDF

Cover design: Lyn Davies Design
Front cover image: Stocksy/Colin Anderson
Bristol University Press uses environmentally responsible print partners.
Printed and bound in Great Britain by CPI Group (UK) Ltd, Croydon, CR0 4YY

FSC
www.fsc.org
MIX
Paper | Supporting responsible forestry
FSC® C013604

Contents

About the Author

Sine N. Just is Professor of Strategic Communication in the Department of Communication and Arts, Roskilde University. She is the principal investigator of Algorithms, Data & Democracy (the ADD-project), a ten-year (2021–2031) interdisciplinary research and outreach project that aims to strengthen digital democracy. A rhetorician by training, Sine remains convinced that persuasion may hold democracy together by tearing it apart.

Acknowledgements

The plan was to ski, write, repeat. It was the spring of 2020, and my family and I had just landed in Boulder, Colorado, with high hopes for our sabbatical. And then … instead of writing a book, I spent most of my time home schooling, internet shopping and doom scrolling. But I also found time to develop a proposal for a thematic call for research on 'Algorithms, Data and Democracy', which became the ADD-project.

If global events had played out differently, this would have been a very different book, not least because it would have been written into (and about) a very different world, but also, and importantly, because it would not have benefited from my many conversations with collaborators in the ADD-project. I am thankful for the vision and support of the Villum Foundation and the Velux Foundation, which issued the call and saw potential in our proposal. And I am happy for the opportunity to work so closely with so many brilliant people – your different ideas and perspectives inspire controversial encounters galore.

In the course of writing, there's a lot of thinking and talking; colleagues at Roskilde University and beyond have been crucial to the process of turning thoughts into words. Members of the research group on AI and Datafied Communication and of the Centre for Digital Citizenship, thank you for all your constructive feedback. Friends and family, thank you for your immeasurable patience.

There comes a time in any young text's life when it must venture into the world and meet readers. Earlier and later versions of this book, in parts and whole, have encountered thoughtful and considerate readers at every turn: Jonas, I hear your scepticism and hope I've addressed at least some of it. Patrick, thank you for your generosity and enthusiasm. Karen, I can't even begin to express how much your willingness to think with *and* against the grain of the text has meant to me.

Extending the 'writing is a journey' analogy, meeting a publisher is maybe the most daunting step. In this case, however, it has also been the most rewarding. Paul, Ellen and team Bristol University Press, thank you for all your help and support. From our first encounter onwards, you have been the best travel guides any author could wish for.

Sara, the opportunity to learn with you has kept me going through it all. I will be forever grateful for all that we are and can do, together – not least, the giving shape to/taking shape from three digital natives and trying to keep it all from blowing up. Mathias, Joseph and Noël, I hope you like them dank memes. Seriously, though, I love you.

CJ, thanks for all the walks. And Ruby and JJ, yes, I know, cats rule.

Finally, I have found that writing a book requires *a lot* of espresso, literal and spiritual. I am thankful for every gift of either or both.

'Controversy', as Yoko Ono says, 'is part of the nature of art and creativity.'

1

The Closing of the Rhetorical Mind

'Can democracy survive the polycrisis?' George Soros asked in June of 2023, listing war, artificial intelligence (AI) and climate change as the three issues that threaten democracy most urgently. With this book, I suggest that the stifling of public debate poses a further challenge to democracy – one that must be resolved in order for democratic societies to become able to deal with their other grand challenges in, well, democratic ways. Public debate, I argue, is being squeezed from two sides: polarization and personalization. These developments have a common source, namely, the algorithmic organization of public debate. Further, polarization and personalization have similar consequences, each adding to citizens' inability to productively encounter opinions that are different from their own and, hence, engage in controversy as a means of reaching democratic decisions.

The first tendency – polarization – is often posited as a significant threat to democracy in its own right (Klein, 2020; Arbatli and Rosenberg, 2021; Kolbert, 2021). Here, polarization may be defined, simply, as the process of antagonizing opposed groups and contrasting viewpoints, which leads the representatives of different positions to move further and further apart. Such deepening rifts in opinions, in turn, are spurred on by – and add further impetus to – dynamics of deepening socioeconomic inequalities (Stiglitz, 2023), leading to the erosion of political as well as interpersonal trust to the detriment of democratic institutions (Theiss-Morse, Barton and Wagner, 2015; Gerbaudo, 2021). While some of the most spectacular examples of polarization and its dire societal consequences emanate from the Anglo-Saxon context (think Brexit, think Trump), the drift is global (Carothers and O'Donohue, 2019). Similarly, polarization is often associated with the rise of right-wing populism and/or nationalism, but empirical research shows that the liberal left is just as biased towards the conservative right as vice versa (Hutson, 2017). Today, whatever your (geographical and/or political) position, you are likely to mostly encounter opinions with which you already agree – and when you do encounter views that you disagree with, it is mostly to confirm how wrong-headed they are (Geschke, Lorenz and Holtz, 2019; Pew Research Center, 2022). This is the first and most basic sense in which personalization and polarization are related.

The tendency to mostly discuss contentious issues with people you already agree with is, of course, not new. Rather, it resonates with Kenneth Burke's (1969) neoclassical idea that identification precedes persuasion. And it affects me as much as it does you. This book is therefore likely to confirm much of what we already accept, you and I, but I hope it will also challenge our opinions and enable us to form new ones along the way.

By invoking our commonality as well as the prospect of a joint learning process, I am trying to make a connection where, in principle, there is none, hoping this might make you amenable to my message. But let's be honest, you could be anyone; I am just assuming that you're not. Thereby, I can feign personalization of a message that is, by definition, standardized. Writing, as Plato complained in *Phaedrus* (1952), does not know who it speaks to; in fact, writing does not 'speak', it cannot answer and it cannot adapt to its interlocutor (see also Derrida, 1981). Still, I am trying to make this text about you, the one person reading right now, in order to become able to persuade you.

Communicators, whether classical rhetoricians or modern marketers, have always sought to personalize their messages, adapting them to their audiences, whether real or imagined (Brown et al, 2018). After digitalization, such identification does not have to be based on guesswork; instead, communication can be personalized down to the level of the individual, based on what your data trail says about your behaviours and preferences (Lee and Cho, 2020). If this were a social media post rather than a book, I would not have to guess who you are or hope to reach you; instead, I could order as many exposures to as many people 'like you' as I could afford. With digital developments, actual personalization is becoming the communicative norm rather than an (unattainable, even if fakeable) ideal.

As such, the political tendency of polarization is accompanied by a communicative tendency of personalization (Keijzer and Mäs, 2022; Zhang, Lin and Dutton, 2022). The messages we encounter are increasingly tailored to our individual tastes – or, rather, the patterns that can be algorithmically induced from our search histories, social media connections, online consumption patterns … the data we produce with every step we take, every click we make.

While clearly connected, the relationship between digital personalization and political polarization is not straightforward (Cho et al, 2020). The more personalized is not necessarily the more polarized, nor does the reverse inference hold. Still, whenever you engage in digital communicative processes, you are likely to encounter more of what you are already interested in and/or agree with (whether you like cat videos, foreign news, quantum physics, all of the above or something else entirely). This diminishes the variety of topics and views you are likely to accept as relevant to you and/or legitimate positions to hold. Thus, what is sometimes termed 'the Overton window', the range of opinions that can be entertained in a polity, is stretched at the collective level (polarization). However, at the same time, the range of viewpoints that may come to hold sway at the individual level is diminished (personalization) (Vo, 2020). And the space in between is emptied out.

At the personal level, digital technologies invite you to follow 'the path of least resistance', offering a state of 'sweet consensus', of agreement with people who also like what you like. This explains why different viewpoints increasingly reach you via the people you already agree with. And it leads to 'nasty conflict', presentations of different opinions in less-than-optimal terms, which is another dominant feature of current communicative encounters. This does not mean you are confined to an echo chamber or a filter bubble – in fact, the empirical existence of these phenomena seems much less pronounced than their conceptual hype (Terren and Berge-Bravo, 2021). Rather, the processes of polarization and personalization exist within a broader media ecology that gives you access – and sometimes exposure – to many more positions than the ones you agree with and their opposites. However, it does mean you will find it increasingly difficult to engage with all of these subtler differences. It means you are less and less likely to witness and participate in public debate over contentious issues.

Technologies of persuasion

Public debate, as I understand it here, is an increasingly digital affair; this digitalization of arenas of exchange of opinions, I argue, means the processes of such exchange are becoming organized along the lines of polarization, on the one hand, and personalization, on the other. This book will show that, together, these two tendencies transform processes of public meaning formation, pushing our collective understanding of matters of public concern further and further into the realms of either the consensual or the conflicting and emptying the space of what used to be thought of as 'the marketplace of ideas' – that elusive arena for the public exchange of opinions. This, I believe, changes how we think of and engage in persuasion, the process of making up our own minds and of seeking to influence the minds of others. Persuasion, in essence, is meaning formation; it is *reasoning with a motive*. How we organize the two – reasons and motives, facts and feelings – in relation to each other is reconfigured by digitalization, changing the conditions of possibility for meaning formation, both individual and collective.

Arguably, where the printing press was all about making things public, digitalization heralds a new intimacy – a personalization of persuasion, if you like (Peters, 1999), or the turning of the public sphere into a private one (Papacharissi, 2010). As we replace broadsheets with small screens, we become unaccustomed to engagement with opinions that are different from our own, and as we become increasingly unlikely to witness public exchanges of opinion, we also become less and less likely to participate in such exchanges. Today, communicative processes are not driven by differences of opinion, but are instead fuelled by showers of love or shouts of abuse.

This is not to say that communication used to be rational and now it is emotional; on the contrary, the argument I am making is premised on the belief that feeling comes – and always came – first. That affect is not the opposite of reason, but its prerequisite (Massumi, 1995). The change we are witnessing, then, is not from communicative processes that were organized rationally to new affectively organized forms of communication; it is a shift in how affect organizes reason. Today, affective drivers of meaning formation intensify similarities centripetally and differences centrifugally, pushing 'likeminded friends' ever closer together and driving people who might otherwise 'agree to disagree' apart. As such, digital communication only follows the first half of the proverb that admonishes you to keep your friends close but your enemies closer; online, friends are kept in close circuit, but enemies are hurled away, making persuasive encounters between people who do not already agree increasingly difficult and rare.

This does not mean that persuasion no longer occurs, but the communication of reasons and motives is becoming the covert work of algorithmic curation, which is tacitly accepted, whereas overt attempts at persuasion are more and more frequently denounced. As 'brokers of information in a digital economy', Jessica Reyman (2018, p 114) argues, 'algorithms do not produce neutral output based on purely objective calculations'. Instead, algorithms are productive of – and run on – new logics, new means of (re)producing (what counts as) knowledge, 'built on specific presumptions about what knowledge is and how one should identify its most relevant components' (Gillespie, 2014, p 168). As such, algorithms are inherently rhetorical; they are shaped by and give shape to distinct forms of persuasion. This is not a unique feature of digital technologies; on the contrary, all technologies organize (Beyes et al, 2022) – and, hence, are persuasive in the sense of selecting some aspects of reality (reasons) and providing the manipulation of these aspects with a sense of purpose (motive). The contentious issue here is not whether algorithms persuade, but *how* they persuade – how they make meaningful connections between reasons and motives.

In answering this 'how', the first thing to notice is that, when organizing processes of meaning formation (that is, when enabling digital communication), algorithms invite political polarization accompanied by cultural consensus; the congregation of likeminded people around the things they like and in opposition to people who like other things. A process that has been termed 'affective polarization' – clustering around views to which one is emotionally attached and against those positions one feels negatively about (Törnberg et al, 2021). In this climate, disagreement gets crowded out or even deplored. As Anand Giridharadas (2022) argues, we are witnessing a 'war on persuasion', a resistance to – and dislike for – communication that openly attempts to 'win hearts and minds', as the saying goes.

This book will examine the entanglements between social processes that encourage 'unpersuadability' and concomitant technological developments towards 'automated persuasion'. I argue that the sociotechnical tendencies of resistance to explicit communicative attempts at persuasion combined with openness to the implicit persuasion of digital technologies leads to a 'closing of the rhetorical mind', an inability to listen to – and produce – persuasive messages. With the allusion to Allan Bloom's (1987) notion of the closing of the American mind, I invoke a deep concern about a general loss of the ability to think critically and reason independently. However, let me hurry to position myself in opposition to Bloom's analysis of the cause of and cure to this problem.

Suggesting that 'liberal education' corrupts young people's minds, Bloom offered an early contribution to the so-called culture wars, the currently raging iteration of which we shall visit shortly. In a later version of Bloom's argument, Greg Lukianoff and Jonathan Haidt (2018) claim that the 'American mind' is being pampered to death; the mind is closed because it is coddled. Contra Bloom as well as Lukianoff and Haidt, I posit that 'absolutism', not 'relativism' is the bane of critical thinking. Today (and in 1987 as well as 2018 for that matter), individual minds (whether young or old) and social collectives are not threatened by the plurality of ideas or by the notion that several ideas might possibly be true at the same time. Further, the fact that what is true or right could change over time does not threaten individuals or the collectives they are part of. Nor is it threatening that some people might, once in a while, ask other people to express their differences more gently. Rather, the greatest threat – at present as well as in the past – is that of (neo)authoritarianism (Giroux, 2022).

Democracy is under threat because we are losing the ability to persuade and be persuaded. This is not an argument against 'snowflakes' who are said to be too frail to bear disagreement, nor is it a reproach of 'cancel culture' – the accusation that someone is silenced or shunned simply for expressing a different view. Rather than positioning myself within the framing employed by either side of the culture wars, I am suggesting that the rhetoric of culture wars per se is supported and perpetuated by digital technologies, and my critique is directed at this sociotechnical configuration. The current entanglements of social norms and technological action opportunities stifle our ability to persuade and be persuaded, reducing our capacity to organize for 'the open society'.

Replacing the historical determinism, which Karl Popper (1994 [1945]) saw as the greatest enemy of liberal democracies, the threat today comes from technological determinism and its cultural effects that, in placing responsibility for societal progress outside of the realm of human agents, hold many of the same limitations – and many of the same risks – as those indicated by Popper. Given rapidly developing digital technologies, including the current hype around generative AI (this book has, or so it feels like, been written in a race with ever-better iterations of large language models that threaten/promise to take over human writing), it is (once again) becoming fashionable to think of technologies as either the solution to or the cause of all our problems.

In seeking to understand and transform current developments, we must resist all types of determinism and all singularities. Technologies do not shape human societies singlehandedly, nor are humans in full control of the technologies they use. Rather, processes of social and technological change are inextricable entangled, always sociotechnical, never just one or the other (Orlikowski, 2007). Furthermore, such entanglement is characteristic of all human societies (Beyes et al, 2022). Throughout history, how we organize has been significantly shaped by the technological developments of the day – just as all societies have developed tools to fit their needs. All technologies, in this sense, are technologies of persuasion; they shape how we understand and interact with the world. Some technologies just make that process more apparent than others.

Today, humans and digital technologies have become inseparable, acting together while obscuring the interaction. Digital technologies have become so ubiquitous that their mode of persuasion now seems neutral. As the digitalization of communication comes to be taken for granted, we have become blind to the contingencies of digital communication and unable to imagine that and how we might communicate otherwise. To become able to organize communication differently, to establish sociotechnical relationships that invite debate, we need tools and practices that wear their persuasive intent on their sleeves, as it were. We need technologies that show us the workings of their persuasive processes rather than pretend there is nothing to see here. To counter the narrowing and tightening of the available spaces for and means of public debate, we need to reinstate the value of persuasion and to improve persuasive processes. We must 'make disagreement good again'.

Ultimately, my argument is that we must re-activate disagreement for the greater public good – and I aim to suggest how that might be done. To this end, the book begins with a diagnosis of the trouble with disagreement, seeking to build new possibilities for public debate out of its ruins. While the diagnosis and the suggested cure are all about sociotechnical entanglements, let us briefly accept an analytical distinction between the social and the technical. In what follows, I attend, first, to the social and, second, to the technical dimensions of today's dominant communicative trends.

We don't talk anymore

One place to start looking for new beginnings is in the ruins of conversation. According to Paula Marantz Cohen (2023), we hardly talk anymore, 'talk' being defined as engaging in deep, meandering dialogues that might take us places we did not expect to go – talk, understood as participation in conversation that might change us, as we expose ourselves to the uncertainty involved in being open to each other. Such conversation, Cohen suggests, is like falling in love, and it is, she argues, a despairingly rare experience these days. The dual dynamics of polarization and personalization have come in its way, leaving us ever more comfortable with repeating established positions and increasingly reluctant to engage with the ambiguity of trying to figure things out together with people whose views we do not already know and share.

One immediate objection to this diagnosis comes to mind: under the current conditions of digitalization, we do not talk less nor are social connections scarce. On the contrary, in 'the Information Age' chatter flows freely and relations are made aplenty as we (re)connect and interact with friends and family online, meet with strangers in chatrooms and groups and engage in broader processes of public exchange on social media. The problem, Cohen argues, is that digital interactions are poor substitutes for conversation; they are not 'safe' for the expression of disagreements or the manifestation of frictions that we might cultivate in face-to-face interaction with people we trust. Instead, they degenerate into hostility from which people are likely to withdraw or they turn into spectacles, through which people may scroll with little commitment and even less effort, leaving them (people and spectacle) fundamentally unchanged.

I agree with Cohen here, but I disagree with the pull towards consensus that is inherent in her concept of conversation. As long as agreement, however thinly defined, continues to be the underlying motive of communication, we cannot resist the pull towards the things we like and the people who are like us. Thus, we can only open ourselves to difference if we learn to resist the temptation to be persuaded by someone just because we identify with them – and, concomitantly, allow for the possibility that we might be persuaded by people who are different from us.

We should, indeed, talk more, but we will not be able to do so, lest we embrace difference and posit disagreement as both the beginning and end of communicative encounters. Conversation, in other words, while clearly resisting polarization, risks becoming but a cover for staying within our current comfort zones, our circles of sweet consensus, whether we find them in seminar rooms or group chats. Conversation may thrive on friction, but if the comfort of community is threatened by disagreement, difference is often the first to go.

Another turn of the culture wars

To illustrate both the ways in which we don't talk anymore and why a conversational model for bringing talk back will not serve us well, let us consider a particularly virulent manifestation of the culture wars as they are currently being waged: the backlash against LGBTQIA+ rights generally and the rights of transgender people more specifically. To begin, we may define 'culture wars' as conflicts between two groups with different beliefs and values that align in such a way as to contribute to societal polarization. The phenomenon includes a turn from political disagreement to 'identity politics', another loaded term, which I use here to describe how views on political matters are thought – or brought – to align with sociocultural positions (Sotirakopoulos, 2021).

Polarizing debates over issues like civil rights, reproductive rights, gay rights, individualism and elitism have defined the US sociopolitical context since at least the 1960s (Thomson, 2010). And with the rise of the 'manosphere', the digital meeting space for all manner of anti-feminist backlash (Nagle, 2017; Ashcraft, 2022), the culture wars and their polarizing consequences have gone online, spreading globally through the internet (Cosentino, 2020). Today, no matter who you are or where you are, if you happen to voice an opinion about any of the contentious issues of the day, you had best be prepared to defend it.

Currently, trans rights – for example, in terms of legal recognition and access to medical treatment – are among the most hotly contested topics around which the culture wars are waged (Castle, 2019). This development follows an extended period of perceived progress across the LGBTQIA+ spectrum. It was a period of decriminalization; for instance, homosexuality became de facto legal in Denmark in 1912, but even so and to this day remains illegal in 64 countries around the world. It was a time for claiming civil and legal rights; most famously, perhaps, the right to same-sex civil unions, pioneered in Denmark in 1989, with full parity for same-sex couples achieved in this national context in 2012. In the US, same-sex marriage has been constitutionally protected since 2015, through a narrow Supreme Court ruling that has not (yet) been overturned. And the recognition of legal rights led to societal mainstreaming, as might, for instance, be indicated by the extent to which nonheterosexual public displays of affection have become unspectacular. Still, in Denmark, a reputedly 'gay-friendly' place (as per the examples given here), 45 per cent of LGBTQIA+ identifying people report that they check their behaviour in public (Wøldike, 2022). What some may perceive as 'too much' mainstreaming, then, is in fact 'too little' for those most directly affected by the negative outcomes of not being mainstream 'enough'.

Given the progress, even if (very) 'slow going' (see Villadsen, 2020), a divide between those who are supportive of LGBTQIA+ rights and those who feel threatened by them has emerged. The backlash is global, perhaps pioneered by Russia but extending in all directions and spreading across ideologies. As we might expect, the anti-trans backlash aligns well with the populist right – a fact that Donald Trump, for instance, has realized and incorporated in his re-election campaign (Seitz-Wald and Yurcaba, 2023). Still, a lot of heavy lifting for the anti-trans agenda is done by a group of alleged feminists who, confusingly, self-identify as progressives while, it seems, being mostly concerned with public bathroom etiquette (Jones and Slater, 2020). For instance, J.K. Rowling, she of *Harry Potter* fame, has seen fit to use her massive public platform to voice opinions that she insists are not transphobic, but that nevertheless align her with contentious arguments about 'the erasure of women' (Romano, 2023).

Aided by such efforts of the proverbial friends that mean real progressives will never need enemies, the backlash has, indeed, grown most fiercely in the fertile ground of the populist right in the US. In April 2022, Governor Ron DeSantis of Florida was among the first to introduce anti-LGBTQIA+ legislation, and as of the spring of 2024 no fewer than 479 additional bills are being debated, decided or deployed across 41 of the 50 states (ACLU, 2024).

As Masha Gessen, a staff writer for the *New Yorker* and outspoken LGBTQIA+ activist, argues in an interview, conducted by their editor in chief, 'all these bills are about signalling, and what they're signalling is the essence of past-oriented politics. It's a really convenient signal because some of the most recent and most rapid social change concerns L.G.B.T. rights in general, and trans rights and trans visibility in particular' (Remnick, 2023). However, the legislative initiatives are not just about signalling; they are also about silencing. Many follow the pattern of Florida's initial legislation, which has been labelled the 'don't say gay' or 'anti-grooming' bill and prohibits public schools from teaching kids about sex and gender identity in ways that are not 'age-appropriate'.

The underlying assumption of such legislation is that the very visibility of alternatives to heterosexual and cisgender identities will somehow be persuasive, and that individuals who embody such visibility are actively 'recruiting' or 'grooming' kids for their 'lifestyle'. One might quip that if exposure equalled persuasion, none of us would identify as anything but straight in the first place. Still, this line of reasoning is, in fact, not totally alien to the pro-LGBTQIA+ 'camp', as is evident from the oft-repeated phrase that 'if you can't see it, you can't be it'. Thus, it may be true that seeing difference makes it easier for someone to identify as different, but that does not mean we are somehow the prey of the differences to which we are exposed. Again, the norm of straightness is what most of us are exposed to the most, but those of us who find ourselves to be different from this norm have somehow managed to resist its lure. Thus, the point should be to present more differences, not fewer, as being exposed to difference seems to protect us from indoctrination rather than cause it.

However, such diversity is exactly what the debate leaves out and makes impossible. And this has the further consequence of silencing differences within the LGBTQIA+ movement(s). As the social commentator Natalie Wynn, aka ContraPoints (2023), says in a video exposé of past and present anti-LGBT legislation, there are many issues that could be discussed (she mentions the age limit for puberty blockers and the question of transwomen in elite sports), but in the current climate raising such issues is more likely to lead to 'cancelling' within the movement than to sustained conversation. Thus, the positions 'for' and 'against' trans rights become increasingly entrenched, and 'both sides' become increasingly resistant to nuanced arguments. Yet, this resistance does not lead to engagement, but instead to avoidance, as it stems from the belief that the mind is meek and should be protected from opinions that it does not already hold, lest it be persuaded by them. The entrenchment of opinions and wariness of persuasion is deeply problematic, from a political as well as a communicative perspective.

Here, it might help if we talked more. But in such conversations, we would have to learn to centre our differences and disagreements as an antidote to the sweet consensus that accompanies nasty conflict. Disagreement, as I invoke it here, is inspired by the example set by ContraPoints, whose very name speaks to the pleasures of the contrary, the delights in pointing to other options, even as she is (often painfully and painstakingly) aware of the vulnerabilities involved in promoting 'the weaker position'. Thus, I am inspired by ContraPoints' activist practice and pedagogical patience, and throughout the book I will seek out and seek to amplify examples of sustained disagreement while also attending to the many situations in which disagreement disappears or breaks down.

In so doing, I take my conceptual bearings from, among others, Jacques Rancière's (2004) work on disagreement, which he defines as involving issues of mutual (lack of) understanding. Disagreement, Rancière suggests, is only secondarily about differences in opinion and primarily about who can speak and be heard; about who has a voice in society. In this conceptualization, voicing difference and recognizing different views is by default a societal and democratic good – a resource we should cultivate rather than a weed to be removed. Today, however, rooting out difference seems to be the dominant activity, which indicates that the alternative to polarization is not conversational consensus seeking or even conversations that are attentive to difference. It would have to be something wilier and more unruly.

We should indeed talk more, but we should talk about what makes us different from each other rather than about the things that make us the same. We should talk in ways that give voice to difference and open up spaces for different voices, avoiding those conversational forms (and norms) that risk becoming univocal over time while being clear about the motives that inform our own and others' conversations. Hence, we should promote open-ended conversations that do not give up on the possibility of persuasion, but instead recognize that persuasion is a two-way street. In order to become able to not only imagine but also practise talk differently, we must understand how current social tendencies correlate with technological developments.

I miss my pre-internet brain

The societal developments that may be lumped together under the headlines of political polarization and social consensus seeking are accompanied – enabled, even – by technological developments towards communicative personalization, which may be understood, more broadly, as having to do with digitalization. Reacting to the digital transformation of society – and to promote a club night of collective internet surfing – Douglas Coupland, who is perhaps most famous for dubbing his (and my) generation 'X', designed a series of posters that he saw as specific to the 'now' of 2012. Among many gems, the dictum 'I miss my pre-internet brain' stands out. In 2018 Coupland returned to this statement, now musing that 'he no longer remembers his pre-internet brain' and considering the consequences of that state of affairs. He lists three main trends or areas of concern (Coupland, 2018): 'The online world has vastly outpaced our ability to create political stability'; 'time now registers in our brain with online data intake and the microscopic dopamine hits it generates'; and 'we are, bored, bored, bored. Next!' With these three indicators of change, Coupland points to shifts in individual and collective experiences of cognitive and social processes; shifts towards the abundant, the immediate and the ephemeral that are underwritten by the infrastructure of the internet generally and its building blocks of data and algorithms more specifically.

By 2018, the digital organization of social reality had fundamentally shifted how we access and perceive information, how we circulate and understand opinions, and how we make the world intersubjectively meaningful. Today, communication is, to a very large degree, defined as and by the algorithmic organization of data (Holtzhausen, 2016). And it is this process that sustains what I refer to as covert – or, perhaps more precisely, inconspicuous or inadvertent – persuasion. In the autumn of 2021, the public was given a rare glimpse 'under the hood' of this process when Facebook whistleblower Frances Haugen revealed how the social media platform, through its algorithms that are optimized for maximum engagement, 'promotes hate speech, damages democracy and is even "tearing our societies apart"' (Bateman, 2021). If and when polarizing messages put forward in inflammatory words are more attention grabbing than, say, nuanced arguments or appeals to peaceful cohabitation, then the former will be prioritized over the latter and promoted in users' feeds, leading to a vicious cycle of more and more engagement with more and more socially damaging and individually hurtful content (Haugen also revealed that Meta, or Facebook as the corporation was still called at the time, knew of the strain that Instagram, owned by Meta/Facebook, puts on teenage girls' mental health).

Inspired by Claudia Aradau and Tobias Blanke (2022), I will explore the *algorithmic reason* that underwrites these processes. And coupling the notion of affective polarization with the process of datafication, I coin the concept of *datafied affect* to indicate how feelings are now organized algorithmically – circulated and intensified in and as data. I will develop these two ideas in Chapters 3 and 5, respectively. For now, the point is that saying 'circulation of datafied affect is organized by algorithmic reason' is a conceptually informed and generalized way of saying 'Facebook profits off hate' (Bateman, 2021), 'YouTube profits off radicalization' (Roose, 2019) or 'Google profits off targeting' (Graham and Elias, 2021). It is an indicator of how digital business models that depend on algorithms to deliver users to advertisers are likely to have democratically adverse effects, as the algorithmic curation of content based on users' data trails funnels individuals into personalized information and interaction flows, bringing them deeper and deeper into more of what they already like – or what users like them also liked.

Such tendencies may make anyone who ever had a pre-internet brain miss it dearly, but as Coupland asserts, there really is no going back, no recovery of what has already been forgotten. Social realities and individual experiences have become thoroughly digitalized and are therefore algorithmically organized as data flows. If we begin from this fact, the question is not how to undo digitalization, but how to do it differently. When Big Tech corporations set the rules of digital engagement, they organize for maximum profit with minimal effort, a process Cory Doctorow (2023) has aptly labelled the 'enshittification' of social media. Through this process, users become the commodities off which tech companies profit, as I will detail in Chapter 2.

Accepting digitalization as the basic condition of possibility does not mean that we should give in to technological determinism and accept the demise of the open society, but it does mean that the task of changing the currently dominant sociotechnical configuration of public debate will not be easy. In what remains of this introduction, I will offer a first preview of how I propose we may nevertheless make some headway.

Break the internet

Digitalization adds personalization to polarization, viscerally turning online interaction towards vitriol – or, as Hsu (2023) writes in an article that attends to this process in terms of the dwindling societal conversation: 'It's telling that social-media platforms, like Twitter [now X], characterize themselves as serving a public conversation, and yet the presence of an audience turns online conversations into performances. A politician today is more likely to dunk on some random hecklers on Twitter than to court them.'

In combination, it seems, current social tendencies and technological developments are bringing out the worst in each other. Yet, before despairing at the prospect of such negatively reinforcing sociotechnical entanglements, let us remember Melvin Kranzberg's (1986) dictum that technologies in themselves are neither good or bad, and nor are they ever neutral. We may specify this 'law' in relation to digitalization and democracy, saying that digital technologies are not inherently detrimental to or supportive of democracy, nor can they avoid shaping political institutions and civil societies. The question is how we, as individuals and collectives, shape and are shaped by technologies – and how we use them to potentially reshape societies and technologies alike.

In this book, I will focus more on end users' experiences of and with technology than on the initial shaping of those technologies. I understand digital platforms as infrastructures for human interaction and begin with the banal observation that we cannot 'break the internet' (despite the allure of the phrase). Along with the 'pre-internet brain', a pre-internet society has been all but forgotten – or, rather, it has become irrecuperable; the option of dialling back on already accomplished digital developments is no longer available.

Still, the prospect of 'breaking the internet' points to the potentiality of using digital technologies in ways that differ from their dominant invitations, hinting at how we might do digitalization differently. Consider what the phrase means; you break the internet by creating content that is so irresistibly attractive that its circulation intensifies beyond the capacity of the internet, thereby 'breaking' it. In this sense, the phrase may be purely metaphorical, as when in 2014 *Paper* magazine used it as the caption for a cover photo of Kim Kardashian; here, it was only an invitation to circulation, a hope that the picture would go viral (as, indeed, it did, but with no consequences for the digital infrastructure that supported this process). However, the original meaning of the phrase was literal, harking back to a time when internet connections – dial-up modems, for instance – were more fragile, tangible and breakable (this was back in the 1990s when people could still remember their – and some even still had – pre-internet brains). Between the literal and the metaphorical meaning of 'breaking the internet' is a space of online havoc, like that characteristic of Anonymous, the hacktivist collective that seems to promote social justice 'for the lulz' (see, for example, Dobusch and Schoeneborn, 2015), or of the K-pop fans who rallied on TikTok to coordinate a buy-up of tickets for a Trump re-election rally in June 2020, causing some embarrassment to Trump's campaign staff who had been boasting that the event was sold out, but found a half-empty stadium (Lorenz, Browning and Frenkel, 2020).

Some of the offspring of what might broadly be referred to as 'internet culture' is purely destructive – or toxic, as seems to be a term of the moment (Beran, 2019). However, the examples given here also indicate how 'breaking the internet' can be a creatively shaping force. And it shows how the practice itself is made possible by the existence of the internet and is shaped by the available invitations to online (inter)action. As such, breaking the internet is plastic in all the three senses of the word, as highlighted by Catherine Malabou (2010, p 15) in her proposition that plasticity presents 'itself as the best suited and most eloquent motor scheme for our time' – as the figure through which we may think about processes of giving shape to, taking shape from and exploding the shape of sociotechnical configurations.

Plasticity, in Malabou's conception, is a thoroughly material form that simultaneously contains and supplants itself: '*There is no exceeding of form that does not assume the plasticity of form and hence its convertibility*' (Malabou, 2010, p 46, emphasis in original). As such, we may use the concept of plasticity as a driver for (explaining) stability and change; giving and taking shape are thoroughly relational processes that, simultaneously, make each other (im)possible. Paradoxically, change emerges from within stability (and stability from within change), whether by 'blowing up' things and rearranging the pieces or by subtly shaping that which we are shaped by through 'repetitions with a difference'.

In the book's concluding chapter, I will turn to the examination of such plastic practices, searching out modes of discussion that are not already too entrenched in dominant developments to obstruct how a discussion of those developments can be had. More specifically, I will be looking for communicative practices of digital organization that enable us to talk about how algorithms and data are shaping societies, and to change how algorithms and data are shaping societal debate. For now, let's consider how that links to making disagreement good again.

Make disagreement good again

As already mentioned, I do not – and cannot – know who you are. However, I imagine that you are interested in the interrelations of digitalization, organization and communication as well as their sociotechnical causes and consequences, especially as regards democratic public debate. Maybe you are already convinced that rhetoric is the key to 'saving democracy', in which case I will not need to 'save persuasion' (Garsten, 2006) to bring you on board, and you can read the following as a confirmation of your views (which will be pleasant, I hope; a bit of sweet consensus never hurts). However, I imagine you as someone who is not entirely convinced, yet not entirely dismissive of the democratizing potentials of persuasion. I am addressing (shaping?) a 'you' that is 'persuasion-curious', and this section aims to persuade you to try it out (that is, to continue reading).

With this description of how I imagine you, I have sought to do the exact thing that I have suggested is becoming dubious – namely, make my persuasive intent clear. I have done so because I think we need more rather than less overt persuasion if we are to turn current developments around and begin the task of reshaping dominant sociotechnical configurations. While we cannot not persuade (and be persuaded), if we do not admit to persuasion, we do indeed become defenceless against it. Thus, we must make persuasion more overt, as pretending there is no persuasion only makes us susceptible to covert persuasive practices of the social, the technical and the sociotechnical kind. This book is devoted to uncovering the persuasiveness of that which pretends to be 'merely' informative, neutral or objective, and to suggesting the democratic potential of alternative, explicitly persuasive practices – that is, practices that aim at dissent rather than consensus and at pluralism rather than polarization.

The field of rhetoric has, throughout history, taken the side of minority positions against the capital T 'Truth' of philosophy; rhetoricians stand with persuasive processes against persuasive outcomes (Conley, 1990; Garsten, 2006). Within rhetoric, the concept of *controversia* (and the practice of controversy) names an approach to persuasion that is particularly attentive to the different sides of a conflict and to the potentiality of disagreement. This understanding of persuasion as a process of finding and considering arguments from 'all sides' of the issue at hand is, I believe, a particularly useful starting point for the task of (re)instating public debate as a source of democratic legitimacy. Chapter 3 will establish current conditions of digital publicness, whereas Chapter 4 is devoted to a recovery and reconceptualization of *controversia*. There, I will offer the classical concept as a better alternative to modern ideals of the public sphere. However, let me preview one aspect of this argument by delineating the classical understanding from present vernacular uses of the term 'controversy' and, instead, connecting it to its use within the field of science and technology studies (STS) and to the concept of agonistics as developed by Chantal Mouffe (2013).

According to dictionary definitions, controversy means 'public disagreement', which need not be a loaded term. Today, however, when someone says that a topic or person is controversial, it is usually not a compliment. Rather, the controversial tends to border on the scandalous and the uncouth; when used in daily speech, it may refer to a somewhat entertaining process, but rarely is it suggested that this process might be democratically beneficial. Chapter 5 will unpack the dismal state, as also introduced in this chapter, in which 'public disagreement' has become subject to fear and loathing.

However, this is not the dominant scholarly view of controversy. STS scholars, in particular, have established controversies as empirical processes in and through which knowledge is socially constructed. Within this field, the focus is, understandably, on scientific controversies, which are demarcated from other topics and issues by five features. Controversies are: (1) made up of all kinds of human and more-than-human actors; (2) social dynamics; (3) reduction-resistant; (4) debated; and (5) conflicts (Venturini, 2010). This characterization enables researchers to map controversies in space and time, noting who is involved, in which ways, in what relations to each other and/or with which outcomes (Venturini and Munk, 2021). Within rhetoric, scientific controversies have also become objects of study, giving rise to the subfield of rhetoric of science, which, in broad terms, focuses on the rhetorical processes that run through and energize controversy maps, as drawn by STS scholars (Gross, 2006).

While I am inspired by both STS and the rhetoric of science, my approach to controversies is broader and oriented towards public disagreements of all kinds. Thus, my understanding is, like that of STS, more descriptive than what is perhaps typical of (rhetorical) studies of public debate. Yet, unlike most STS scholars, I do hold to a normative ideal, namely that of disagreement as advocated by Rancière, where a society's capacity for disagreement is a measure of its plurality as well as its inclusivity. This position aligns well with Mouffe's vision of agonistic politics, and it combines an STS-like affinity for broader descriptive strokes with a rhetorical attention to normative detail. Mouffe defines agonism as a 'struggle between adversaries' who share 'a common allegiance to the democratic principles of "liberty and equality for all", while disagreeing about their interpretation' (2013, p 7).

Following Mouffe, we should focus on both the means and ends of debating, on articulated differences and common underlying assumptions. When attending to the articulation of public debate, the normative measure is not whether interlocutors were able to reach agreement, but, on the contrary, whether they were able to sustain engagement while recognizing differences – that is, were they able to uphold the principles of 'liberty and equality for all'? This is the aim of what I term 'controversial encounters': a debate that never ends. If we are to sustain democracy, adversaries must be willing and able to continue talking – for the very sake of supporting disagreements (and debates about them).

Kiss the specifics

To study and promote controversies is, in part, a conceptual endeavour; it is about theorizing current conditions of public debate and finding conceptual alternatives. Thus, this book's ambition is largely theoretical as I am presenting an interdisciplinary lens for studying controversies. This lens draws together organization and communication studies around issues of digital public debate and posits rhetoric as the point of refraction where the different disciplinary insights both converge and diverge, establishing a combined conceptual framework for the study of disparate empirical processes.

The big argument being made throughout the book is that digital organization of public debate pulls processes of meaning formation in the opposite directions of consensus and conflict, abandoning the types of meaningful encounters with differences of opinion that I term 'controversial'. Based on this diagnosis, I suggest that to save democracy, we need to attune processes of meaning formation to other affective forces than those of polarization and personalization. Leading with affective alternatives, I argue, we can begin the process of reshaping the sociotechnical configuration of public debate, creating the basis for sustained engagements with controversial encounters that are, in fact, both *controversial* and *encounters*.

This conceptual argument about general tendencies finds inspiration in specific empirical instantiations. As such, I do not believe that there are any conceptual models that can explain all of public debate and nor are there any practical steps that will reconfigure it all at once; rather, we must look for and nourish those instances that work – and learn from those that don't. We should follow controversies as they unfold and salvage – promote, even – whatever democratic potential for articulating and continuing relationships of disagreement they might hold.

In advocating that we 'kiss the specifics' (in the words of singer-songwriter Joan as Police Woman), I am inspired by Jack Halberstam (2011), who produces 'low theory' by engaging with processes and phenomena of various kinds. Halberstam shows that it is through such engagement that we can learn from our present realities rather than get caught up in futile attempts at transcendence. Halberstam's own engagements span phenomena of popular culture and queer art, from the animated antics of SpongeBob SquarePants to the alarming artwork of Cabello/Carceller. I follow this heterogeneous practice of engaged theorizing, finding my examples across the spectrum of politics, both low and high, everyday interactions, internet distractions, among activists and reactionaries of the left and the right and generally drawing on the various persuasive processes that caught my attention during the course of writing this book.

The empirical eclecticism and methodological laxness (in fact, you will find no new empirical studies in this text) of this approach to theory building is not only open to criticisms of cherry picking and subjectivity; establishing my conceptual position also has to 'forfeit a degree of nuance', as Laura Mulvey has said of 'the male gaze', a concept she coined to capture a historical moment rather than to develop a lasting theory (Jackson, 2023). In other words, the theoretical claims I will make involve some degree of 'strategic essentializing' (Shome, 1996) – or what Kate Lockwood Harris and Karen Lee Ashcraft (2023), invoking the work of Karen Barad, conceptualize as 'cuts', the process of producing 'slices', whether conceptual or empirical, by separating some entities off from others. Since all networks, in principle, continue indefinitely, deciding which relations to highlight and which to ignore is indeed always a choice (an agential cut, in Barad's terms) – and one that contributes to the configuration of what is studied.

In what follows, I will, as Harris and Ashcraft suggest, seek to widen my conceptual networks, showing not only from where I get my inspiration but also endeavouring to seek out the sources of inspiration of the authors that most inspire me, paying homage to broader flows of ideas and intricate relations of scholarship. Inevitably, however, cuts will be made, some of which will be less conscious and others more arbitrary. Thus, you will notice some cuts and wonder why I have left out a particular framework or concept. Other cuts will appear seamless, as they follow your intellectual tastes and expectations. Either is equally necessary – and equally problematic. Hence, I ask you to bear with me on both counts, hoping you will find that the conceptual choices I have made balance comprehensiveness and consistency.

Similarly, I have made many and often rather severe empirical cuts, selecting examples that fit my conceptual leanings as well as my sociopolitical orientations and preferring theoretical rabbit holes to empirical fields. While I will do my best to not lose sight of the empirics altogether and to present a broad array of controversial matters, including controversies that are geographically dispersed, my outlook is admittedly Western-centric (with a predilection for the US, despite my anchoring in Northern Europe), and my tastes lean towards refutations of the claims that 'the left can't meme' as, for instance, provided by Alexandria Ocasio-Cortez's 'tax the rich'-statement dress, which she wore for the 2021 Met Gala (Villarreal, 2021). Conceptualizing with such examples enables me to present *one* view of the world that, I should emphasize, is exactly that: one of several available perspectives. Throughout the book, I will strive to be clear about my positionality, but I will also seek to make my case as strongly as possible. Thus, I am hopeful that my theoretical concepts and empirical illustrations will hold persuasive appeal beyond readers who already agree with the argument I am seeking to make.

Besides illustrating why we might still find progressive potential in meme culture, Alexandria Ocasio-Cortez's dress (and the furore around it) perfectly summarizes the ways in which digitalization leads to online–offline integration, communicative and organizational convergence and entanglements of production and consumption. These three aspects of digitalization will be the focus of Chapter 2, which establishes the foundation of my argument within (the study of) organizing digital communication.

In Chapter 3 more foundations will be laid as I turn to the question of how digitalization transforms the public sphere, specifying my area of concern as the organization of digital public debate. Here, I argue that algorithmic technologies organize calculated publics, which lend themselves more to spectacular antagonism than to concerted engagement. This is a process that can be illustrated by the brawl between Mark Zuckerberg and Elon Musk, who in 2023 loudly and protractedly challenged each other to a 'cage match' of mixed martial arts, thereby drowning out serious conversations about how they each run their platforms. By the end of 2023, the match seemed to have been 'called off' (Vega, 2023), but for a while my corner of the internet was buzzing with the exciting prospect of an entertaining spectacle disguised as controversy – a spectacle that was both nourished by polarization (are you 'team Zuck' or 'team Elon'?) and amenable to personalization (haven't heard of the proposed cage fight? Well, maybe your tastes are less warped than mine).

Chapter 4 begins building my conceptual alternative to the digital transformation of the public sphere, as established in Chapter 3. Here, I turn to classical rhetoric to recover and redefine the notion of controversial encounters, which is, in Chapters 5 and 6, applied to current configurations of public debate.

Chapter 5 begins where Chapter 3 left off and details the detached entrenchment that results from processes of personalized polarization, which play out as sweet consensus *and* nasty conflict. As already indicated, this is the core of my diagnosis of the current configuration of public debate, constituting what I term the closing of the rhetorical mind.

In Chapter 6 I look to the immediate future, which is fast becoming the present, and discuss how the advent of generative AI is pushing us further in the direction of automated persuasion and exasperating the current situation by making us even less accustomed to overt attempts at persuasion.

Thus, the call to make 'disagreement good again' is becoming ever more urgent, and in Chapter 7 I begin building an inventory of affective alternatives to detached entrenchment. Here, I will suggest how we are already engaging in and with controversies in ways that may help us encounter each other and our differences, opening up debates to continued disagreement rather than narrowing them down to insurmountable conflict or parsimonious consensus.

Finally, Chapter 8 summarizes what has been learned about the closing of the rhetorical mind and offers my combined suggestion for making disagreement good again – it outlines a plan for the rescue of digital democracy, beginning with a reconfiguration of controversial encounters that will enable us to stay in conversation about our disagreements. Just as the book offers but one possible diagnosis of public debate in the age of algorithms, it offers but one potential alternative, one vision for how democratic societies may thrive. Returning to Soros' question of whether democracy can survive the polycrisis, making disagreement good again may not guarantee the resolution of all crises, but without controversial encounters, democracy is already dead – or at least that is my position.

2

Press Play: Organizing Digital Communication

In April 2021 a famous photograph was sold for nearly $500,000. While that is not an unusual going rate at art auctions, this photograph was not your usual work of art, nor was it sold at a traditional auction house. Instead, the picture, a close-up of a child who is standing in front of a burning house and has turned her head to smile knowingly at the camera, forms the basis of the 'Disaster Girl' meme, which exists in innumerable copies and countless versions across the internet. Further, Zoë Roth, the now adult protagonist and owner of the photo, which was taken by her father in 2005, did not sell a print version of it, but the digital original in the form of a nonfungible token (NFT), a unique digital code stored on blockchain. Finally, the transaction took place on Foundation, an auctioning site for digital art where users can create, buy and sell NFTs, using Ether, a cryptocurrency. Although selling the original cannot end circulation of the meme, Roth explained that the sale 'was a way for her to take control over a situation that she has felt powerless over since she was in elementary school' (Fazio, 2021). Also, it enabled her to pay off her student loans and donate to charity.

Empirically, the story of the meme-turned-NFT is illustrative of a cycle of boom and bust, as the sale was supported by a gold rush to the markets for cryptocurrency generally and NFTs specifically, creating financial bubbles that have subsequently burst (Alexander, 2022; Hawkins, 2023). Roth, it seems, got in and out at the right time. Conceptually, the story enables us to engage with three central features of digitalization: the collapse of contextual distinctions between online and offline realities; the convergence of communicative and organizational processes; and the financialization of affective circulation.

In this chapter, I will, first, establish the three features of digitalization, discussing them in relation to Jodi Dean's concept of communicative capitalism. Communicative capitalism, as Dean defines it, is 'a political-economic formation in which there is talk without response, in which the very practices associated with governance by the people consolidate and support the most brutal inequities of corporate-controlled capitalism' (2009, p 24). More specifically, Dean argues:

> In the 21st century, communication plays a fundamental role at the level of production, consumption and the circulation of goods and natural resources. Because of the rise of networked media, informatisation and global communications networks, communication supplies the resource for accumulation, functions as means of accumulation, and works as a tool for accumulation (for mining and processing communicative data). (2019, p 331)

This valuing of communication in economic terms, Dean suggests, leads to its political devaluation. As such, communicative capitalism circulates 'consumerism, personalization, and therapeutization' (Dean, 2009, p 6), offering individualized solutions to collective problems and creating an environment of mutual detachment between 'the people' and the democratic institutions that are meant to serve them. Communicative capitalism, then, is the societal formation in which communication is monetized and depoliticized. There is, so to speak, a lot of talk, but no action – no obligation on political actors to respond to, let alone act on publicly circulated messages, whether of support or opposition. Or, as Dean quotes Giorgio Agamben for stipulating, communication 'is constituted as an autonomous sphere to the extent to which it becomes the essential factor of the production cycle. What hinders communication, therefore, is communicability itself: human beings are being separated by what unites them' (Agamben, 2000, p 115, quoted in Dean, 2005, p 56). The more things are communicated, the less they mean and the more ephemeral they become – or, as Dean poignantly puts it, 'without a body to implement a strategy, all the suggestions in the world are little more than noise, contributions to the circuits of communicative capitalism easily drowned out by outrage and puppies' (2019, p 338). When everything is communication, nothing persuades; communicative capitalism, in short, produces the closing of the rhetorical mind.

The overriding argument of this book owes a great debt to Dean's diagnosis of communicative capitalism. However, my position is, as will become clear in this and subsequent chapters, also different from Dean's insofar as I accept affective circulation as a prerequisite for organization under conditions of digitalization, exploring the agential constraints (opportunities *and* limitations) it presents for economic as well as political action, and for individual aims as well as collective gains. Thus, my argument begins from the assumption that technologies shape the organization of communication just as communicative organizing shapes technologies – an assumption that

I seek to substantiate in the present chapter, thereby laying the ground for the argument that follows.

To do so, I detail the three key features of the digital organization of public debate, returning time and again to the illustrative case of Disaster Girl. And discussing the three features in relation to what Dean (2005, 2009) terms the fantasies of communicative capitalism: the fantasy of abundance, which privileges contribution over meaning; the fantasy of participation, which fetishizes technology; and the fantasy of wholeness, which prevents antagonistic relationships that enable the societal inclusion of difference and, instead, figures the other as a threat that cannot be contained, let alone accepted, but must, instead, be destroyed.

On that basis, I go on to consider the consequences of digitalization for the exercise of individual agency and the concomitant formation of digital subjects. What, in other words, does it mean to perform digital labour? In considering this question, I offer a first indication as to why I believe there is no alternative to affective circulation, establishing the point at which my argument branches off from Dean's position. What we need is not less communication or even communication that is fundamentally different from the current mode of circulation; what we need is for communication to circulate differently. In Chapter 3, I explore why that will not be an easy course of action but is nevertheless the only viable route.

Online–offline integration

The Disaster Girl meme illustrates online–offline integration in the basic and straightforward sense that the meme is based on an offline event that has come to take on an online life of its own – a life which, as Roth puts it, is beyond the control of those involved in its creation. Still, the online circulation of the picture has had real consequences for Roth and her family – consequences that may be somewhat particular to this case but are nevertheless illustrative of a broader tendency for digital content to 'travel' not just across different online platforms but also back into the offline realm from whence the content typically came. The meme is a product of the 'new experience of hypermediated reality' (Floridi, 2015, p 1) for which Luciano Floridi and his collaborators have coined the concept of 'onlife'.

We may understand onlife experiences in terms of mediatization, noting how media generally are defined by their ability to 'mediate' (social) reality. As Stig Hjarvard (2008, p 126) explains, 'the media are at once part of the fabric of society and culture and an independent institution that stands between other cultural and social institutions and coordinates their mutual interaction'; the ability to 'connect' different worlds is a defining feature of all media. What is particular to digitalization, then, is not that we know the world through media, but how we do so. Mediatization, we might say, is not just an epistemological condition of possibility, but an ontological reality. Onlife experiences are not just the sum of online–offline integration, but exist on a distinct experiential level. This level, in Dean's terms, is that of the 'zero institution', which 'has no determinate meaning but instead signifies the presence of meaning' (2005, p 67).

As such, the world itself is increasingly made up of mediated communication. Klaus Bruhn Jensen (2013) suggests that we may capture the duality of experiencing the world though media and media as the world by moving beyond definitions of mediatization that seek to delineate the empirical phenomenon to which the theoretical concept applies. Instead, we can use it as a 'sensitizing concept' that attunes us to three central aspects of mediated reality: social structuration, technological momentum and embedded communication.

First, mediatization is key to unlocking the classical structure-agency impasse since 'communication lends meaning both to structures emanating from the past and to agency shaping the future' (Jensen, 2013, p 214). Thus, media are 'a constitutive component and a necessary condition of social structuration throughout the history of human communication and media technologies' (Jensen, 2013, p 214). The question to ask, anywhere and at all times, is not whether media shape reality, but how they do so.

Second, 'each communication technology is a material resource whose distinctive features help to explain the media institutions and communicative practices that have emerged, or which may emerge in the future' (Jensen, 2013, p 216). The specific materialities of various media invite different enactments of reality, and for digital technologies the invitation is to enact mediated and nonmediated realities in such a way that they become increasingly entangled. In other words, digital media invite communicative (inter)actions that cut across contexts, as digital *affordances* embed the experience of reality in (mediated) communication.

I will develop the concept of affordances gradually as I apply it recurrently throughout the book, adding nuance at each appearance. However, an initial definition may be in order, for which I rely on Taina Bucher and Anne Helmond's (2018) conceptualization of affordances as multilayered and relationally constituted, technologically, socially and communicatively conditioned action possibilities – or invitations to action – that exist independently from yet are dependent on their perception and realization. While the last part of this definition may seem overly cryptic, it merely indicates that it makes a difference whether a feature of technology – say, the option to post on a social media platform or to like other people's posts – is used or not; the feature exists independently of individual usage, but it is unlikely to persist for long if nobody does, in fact, use it. This also indicates how affordances are tied to (digital/media) platforms, embedded within specific sociotechnical infrastructures and moving between them, enabling connections between platforms, as with the hyperlink, or travelling between them as when one platform adopts a feature that first appeared on another.

Moving to the third aspect of mediatization, the embedding of communication in reality (and reality in communication) takes different forms. Communicative resources are embedded in each other when, for instance, 'old' media (newspapers, radio and television) are integrated into 'new' (digital) media. Different contexts of action become embedded, as digital technologies are increasingly integral to people's interactions with each other and with the world. Hence, digital actions have become mundane practices, as is, perhaps, best illustrated by the experience of finding one's way by using a map on a phone. Finally, digital technologies are increasingly embedded in physical phenomena (natural environments as well as the human body). Notably, this final feature takes the form of devices for extracting data and, increasingly, acting on the data source, be it a human body, an animal or plant or a manufactured object (the 'smart' phone, the 'smart' car, the 'smart' home).

Within the general conditions of mediated possibility, hypermediated reality involves the more specific experience of 'context collapse'. This collapse begins with digital technologies' negotiation of traditional spatiotemporal boundaries, enabling people to (re-)enact experiences of 'the here and now' across chronological time and physical space (Gulbrandsen and Just, 2011). From the connection across contexts, it evolves into an inability to keep different social spheres separate. Disaster Girl illustrates these different dimensions of context collapse through the continued and global circulation of the meme beyond the control of the photographer (Roth's father) and the subject (Roth herself). As such, the 'here' and 'now' of onlife experiences are continuously haunted by the 'there' and 'then', creating similar difficulties for different actors. Like Zoë Roth, who cannot control the circulation of Disaster Girl, we all live in a hypermediated reality where any number of communicative messages circulate on the same plane, always available for further (re)discovery, (re)mediation and (re)circulation, pushed along and (re)mixed by individuals, but beyond the control of any one actor.

When moved online, offline actions seemingly happen everywhere and last forever, and organizations and individuals alike must be ever attentive to the coherent construction of their selves, even as it forever eludes them. It may be that, as the classic *New Yorker* cartoon has it, 'on the internet nobody knows you're a dog'. However, that is only true as long as you perform consistently and 'authentically' as your human persona – and in many cases that will involve offline actions, both past and present, which would soon expose you if you were anything other than the human you say you are (Haimson and Hoffmann, 2016).

Beginning with Joshua Meyrowitz's (1985) reinterpretation of Erving Goffman's frontstage-backstage dichotomy for the age of 'electronic media', the erosion of boundaries between public and private (or 'staged' and 'authentic') selves has been the subject of much concern in media and communication studies (Davis and Jurgenson, 2014). What happens to the self when it has no place to hide or rehearse, as it were, but must perpetually perform itself in public? And what happens when one's performance in one spatiotemporal context can never be fully contained to that particular moment in space and time, but is always at risk of flowing over into other situations, despite their nominal spatial and/or temporal distance? For Roth, the answer has been to literally take ownership of Disaster Girl, cashing in on the online fame she never asked for or has had any say in.

For others, the negotiation of 'networked identities' is an ongoing process of maintaining boundaries while establishing connections (Quinn and Papacharissi, 2018). Here, the everyday (offline) experience of being able to foreground certain aspects of one's identity if and as relevant while (more or less) freely downplaying or, indeed, hiding other dimensions and roles is contradicted by the need to manage all nodes of one's (online) network all the time. In other words, online you cannot go to work and be a professional, come home and be a partner and a parent, visit your own parents and foreground those parts of your identity that remain consistent with who you were as a child, then go on to meet up with friends and laugh about it all. While certain nuances and elisions are surely possible, the risk of being called out if you deviate too much from anyone's expectations of you is so great that chances are you end up sticking to a vaguely generic version of you that no one fully recognizes but everyone will find 'mostly harmless'.

This development is particularly thought-provoking as it goes directly against the grain of the equally mundane experience of online spaces as free-for-alls in which you can play with different characters, put on different masks and personae, try out acts and attitudes you'd never dare entertain anywhere else. In this sense, online experiences are both less and more real than the 'real' life, which many of us still suppose happens offline. The problem here is exactly that of context collapse, of not being able to keep the real and the play apart, which can be especially worrisome for those of us who are playing for real and go online to find safe spaces in which to bring to life those dimensions of our identities that it might be difficult, even dangerous, to put on full display anywhere else (Triggs, Møller and Neumeyer, 2021).

In sum, online–offline integration supports what Dean (2005) labels the fantasy of wholeness: the idea that everything is connected and that global unity is Real (in the Lacanian sense of existing outside of symbolic meaning formation). This leads her to diagnose the internet as a zero institution, an empty signifier to which any and all meanings can be attached. As Dean explains:

> Despite the fact that its very architecture (like all directed networks) entails fragmentation into separate spaces, the Internet presents itself as the unity and fullness of the global. Here the global is imagined and realized. More than a means through which communicative capitalism intensifies its hold and produces its world, the Internet functions as a particularly powerful zero institution insofar as it is animated by the fantasy of global unity. (2005, p 67)

The intense online circulation of different opinions notwithstanding, different online experiences cannot be recognized. Rather, each individual comes to imagine that whatever they encounter is the totality of what exists: 'precisely because the global is whatever specific communities or exchanges imagine it to be, anything outside the experience or comprehension of these communities either does not exist or is an inhuman, otherworldly alien threat that must be annihilated' (Dean, 2005, p 69). For this reason Dean (2003) believes that the internet is not a public sphere. As I will discuss in Chapter 3, I take a different view on the potential publicness of connectivity; this view is related to my take on communicative and organizational convergence, the second feature of digitalization.

Communicative and organizational convergence

Context collapse also occurs at the level of digital infrastructures, which Henry Jenkins theorizes in terms of convergence culture. In the digital media ecology, Jenkins argues, content flows across multiple media platforms as different (media) industries collaborate to cater to 'the migratory behavior of media audiences who would go almost anywhere in search of the kinds of entertainment experiences they want' (2006, p 2). More specifically, this flow of content is enabled by the digital affordance of 'spreadability', which 'refers to the potential – both technical and cultural – for audiences to share content for their own purposes, sometimes with the permission of rights holders, sometimes against their wishes' (Jenkins, Ford and Green, 2013, p 3). This point is neatly illustrated by the Disaster Girl meme, which was created and circulated beyond the control of Zoë Roth and her family, and continues to spread uncontrollably after the sale of the NFT. For as long as anyone can find use for it, the meme will keep spreading, moving across contexts and causing them to converge.

Conceptually, the point aligns with what Dean terms the fantasy of participation, which leads people to believe that any and all online contributions matter: 'people believe that their contribution to circulating content is a kind of communicative action. They believe that they are active, maybe even that they are making a difference simply by clicking on a button, adding their name to a petition or commenting on a blog' (2005, p 60).

Dean explains this belief in terms of 'tech fetishism': 'what is driving the Net is the promise of political efficacy, of the enhancement of democracy through citizens' access and use of new communications technologies' (2005, p 62). Here, technology becomes a fetish for political action; that is, it stands in for and acts on behalf of the actively engaged political subject and is filled with the subject's political hopes and aspirations.

The fantasy of participation finds further support in digital infrastructures' overriding memetic logic, which offers up imitation as the appropriate form of participation. While the logic of memes is expressed as repetition, it is neither limited in content nor form (Shifman, 2014). Although a meme often takes the form of a humorously captioned picture, it can carry any message in any modality: a political stance in a TikTok dance, a hashtag of solidarity or adversity, building community and dismissing those not 'in the know' with a simple phrase, like, OK boomer. And while memes work through replication, each repetition incurs the possibility of difference. As a case in point, the Disaster Girl meme is typically used to express knowledge of or complicity in calamity, but the manipulation of the image as well as the adaptation of its accompanying text allows for great variation as to severity (from the death of a spider to the sinking of the *Titanic*) and sentiment (from malevolence to apathy).

The term 'meme' has itself travelled memetically. It was coined by Richard Dawkins in his first book *The Selfish Gene*, which, incidentally, was published in the same year I was born (1976). Yet the 'middle kid' of my household (who was born in 2008) once interrupted my excited rant about writing this book to exclaim (with a hint of respect but mostly forbearance), 'wow, you know what a meme is'. To be clear, the point here is not that kids who were born into a world already populated with digital memes like Disaster Girl are particularly susceptible to Dawkins' controversial position that selfishness is the basis of evolution. If anything, 'the kids' are less likely to accept the premise of selfishness, as they are growing up in a media context that persistently serves up experiences of decentred agency. Thus, the predominant experience of memetic content is that it evolves independently of individual users, organized by the technical specifications of the platforms on which memes circulate rather than by the people who circulate them (Rogers and Giorgi, 2023).

This understanding of memes offers an apt starting point for the further consideration of how communicative and organizational processes converge under conditions of digitalization. Contra Jenkins, we might posit that audiences do not share content for their own purposes, but for the purposes of the content they share. This is also Dean's point when she argues that online participation is a depoliticizing fantasy, 'because the form of our involvement ultimately empowers those it is supposed to resist' (2005, p 61). Again, however, while I share Dean's dismal diagnosis of the current situation,

I disagree that we need to step outside of digitally organized communication if we want to repoliticize communication. Or, rather, seeing as there is no 'outside' of processes of digital communication, whatever changes can be made must come from within these processes.

This raises two questions: who communicates and what is created in and through communication? Traditionally, individual and collective (human) actors have been seen as the subjects of communication, but the approach of communicative constitution of organization (CCO) suggests we turn the arrow of causality around, making (organizational) actors the results of 'activities of mediation' (that is, communication) rather than the progenitors of (speech) acts (Cooren and Taylor, 1997). Since its introduction in the 1990s, CCO scholarship has proliferated and branched out into several schools (Schoeneborn et al, 2014), which, subtler differences aside, share the basic assumptions that communication processes (or 'events', understood as the intermittent results of these processes) should be the basic unit of analysis and that there is nothing 'outside' of communication. In other words, one should not expect to find any causal explanations of organizational forms, but could instead seek to identify the communicative events that constitute organizing, understood as networked configurations in the here and now as well as across space and time (Cooren et al, 2011).

Building on these basics, Blaschke, Schoeneborn and Seidl (2012) suggest conceptualizing organizations as 'networks of communication episodes', centring the issue of how actors in an organizational network constitute that network – as well as their positions in it – through communicative interaction. As the notion of communication episodes indicates, such interaction may be episodic, even singular, but organizations become stabilized if and when communication is repeated. And, conversely, 'while communication episodes collectively contribute to the stabilization and continuous reproduction of the organization as a whole, they are inherently dynamic in nature' (Blaschke, Schoeneborn and Seidl, 2012, p 899). Here, the relations between the members of an organizational network are highlighted; while individual actors can talk organizations into being, such actorhood is not precommunicative either. Rather, the nodes of the network are as communicatively constituted as the network itself.

The notion of communicative constitution may suggest radical epistemological constructivism, but ontologically CCO is firmly committed to the communicative powers of matter and the mattering powers of communication (Ashcraft, Kuhn and Cooren, 2009). Hence, proponents of CCO subscribe to a relational and 'flat' ontology, which puts the social and the material on a par as equally essential to the existence and experience of everything (Cooren, 2018). As such, human beings not only communicate intentionally via media but are also media for the communication of ideologies, institutions, emotions, environments and much more. And communicative processes are shaped by social norms and practices as well as technological, physical and structural conditions of possibility.

Evidently, the convergence of the social and the material – and of organization and communication – is not particular to digitalization, but it is particularly important to the study of organizing digital communication for two reasons. First, as studies of collectives and movements like Anonymous (Dobusch and Schoeneborn, 2015), Occupy Wall Street (Kavada, 2015) and the alt-right (Eddington, 2018) make clear, digital organizations are communicative without remainder. Second, this might suggest that digital organizations are purely social and that 'virtual reality' is not 'really real', but, as Paul Leonardi (2010) warns, following this hunch would be a mistake. We ignore the materiality of the digital at the peril of overlooking the ways in which digital materials condition actions and experiences as well as minimizing the very real consequences that virtual encounters may have.

Digitalization, then, deepens the convergence of communication and organization, further enmeshing the social and the material – and we should pay particular attention to the latter if we want to understand the former. This is by no means a new insight; in fact, it is as old as online interaction itself, as illustrated by Julian Dibbel's (1993) seminal exposé of 'a rape in cyberspace' – an account of the author's experiences with LambdaMOO, an early online meeting place (a MOO is an 'object-oriented' 'multi-user dungeon'), which powerfully reasserts the power of communication.

Indeed, if you can do things with words, as has become an increasingly commonplace assumption since J.L. Austin (1962) laid out the basic premises of speech act theory, why should shifting the context of an utterance from an offline to an online setting diminish the performative power of that utterance? A threat is a threat is a threat and can be experienced as threatening no matter where it is uttered. However, the members of LambdaMOO did not speak of the experience on which Dibbel reports as a threat; instead, they experienced an actual violation. This confirms the depth of online-offline convergence, speaking to the ways in which online experiences not only feed on offline events, but bleed (back) into the offline world, leaving marks on minds and bodies as onlife experiences become the ubiquitous mode of encounter.

Further, it suggests that online participation – whether in the destructive mode of abuse or circulating as other intensities – is not just a fantasy. While I agree with Dean (2005) that we should be wary of 'wildly displaced enthusiasm over the political impact of a specific technological practice' (2005, p 62), the out-and-out rejection of the political potential of any technological practice, I believe, would be equally displaced. That is not to say the political impact of online circulation is good – recalling Kranzberg's dictum, no technology has an inherent valence. What I am saying is that we can change its direction and value.

The financialized circulation of affect

This leads to the third feature of digitalization, and to what Dean terms the fantasy of abundance, which is, in her account, the central reason messages now circulate without engagement: 'The fantasy of abundance covers over the way facts and opinions, images and reactions circulate in a massive stream of content, losing their specificity and merging with and into the data flow. Any given message is thus a contribution to this ever-circulating content' (Dean, 2005, p 58).

The third feature of digitalization, then, follows directly from the second, since distributed digital actions are ripe for monetization. With her notion of 'affective economies', Sara Ahmed (2004) explores the relationship between meaning and money in analogous terms, implying that the exchange of symbolic signs is *like* economic transactions. Thus, 'signs … increase in affective value as an effect of the movement between signs: the more they circulate, the more affective they become, and the more they appear to "contain" affect' (Ahmed, 2004, p 120). In this conception, communication is like capitalism and specific utterances (or affective signs in Ahmed's parlance) are like the currencies of monetary economic systems; their values increase and decrease with the ebbs and flows of their circulation.

In digital circulation, this analogy is transformed into a direct relationship; the more signs circulate, the more valuable they become – in affective *and* economic terms. Digital capitalism, as Dean makes clear, *is* communicative capitalism; economic value is derived directly from the circulation of 'content'. And this circulation is emotional, meaning that communicative capitalism is 'emotional capitalism' (see Illouz, 2007). Again, Disaster Girl is illustrative of this, as it is the vast circulation of the meme that enabled Roth to profit from the process she had hitherto felt 'powerless' over; the existence of the countless 'copies' of the meme became the very reason anyone would want to buy 'the original'. Like other beneficiaries of the 'attention economy' (Davenport and Beck, 2001), Roth was able to cash in on the circulation of a sign to which she happened to have a special claim. For others, like the Uber driver or the Instagram influencer, this is not a one-off compensation, but rather a matter of constantly maintaining one's position in an algorithmically organized order-flow (Rosenblat, 2018). And for the tech companies, profit derives directly from 'the data flow'. I will return to the issue of value extraction from digital labour, but first there is more to be said about the financialization of affect (Konings, 2015).

The economy as a whole is increasingly financialized, meaning that profit is derived directly from trade in financial products (stocks, mortgages and so on) rather than from the exchange of services or goods (van der Zwan, 2014, p 103). Similarly, conditions of digitalization lead affective circulation to intensify itself without reference to anything beyond its own circulation. The circulation of a meme leads to further circulation, just as interest in a stock may lead to more interest. And just as the financialized economy is, in fact, no less real than – or has just as real effects as – the so-called real (or productive) economy (Palley, 2013), so the digital circulation of affect is material precisely because of its virtuality, its ability to form and intensify feelings (Massumi, 2014). Ultimately, affective and economic values become directly exchangeable, one leading to the other in such a way that the sale of Disaster Girl is no different from the trading of any other 'synthetic' financial product.

The link between financialization and affect in the context of digital technologies has not gone unnoticed. For instance, Adam Arvidson (2016) demonstrates how social media like Facebook operate on the same logic as financial derivatives. As such, Facebook can size up and slice up its users in myriad ways and prepare and package their data for purposes of resale – just like financial service providers are constantly inventing new synthetic products whose value is purely speculative, but no less real as long as the speculation holds up. In Arvidson's telling, this is primarily an argument about the financialization of everyday life by technological means; it is now possible to speculate in derivatives of personal data in exactly the same way as one would do with bundles of mortgages and other financial products. In both cases, what is actually for sale is people's personal and emotional lives.

However, the argument also works at the systemic level where personal experiences merge into the financialization of social imaginaries (Haiven, 2011), transforming society as such into an object of speculation – or, as Nancy Fraser (2022) puts it, shaping society in the image of and for the purposes of financialized capitalism. The resulting societal shape, Fraser argues, is 'one that authorizes an officially designated economy to pile up monetized value for investors and owners, while devouring the non-economized wealth of everyone else' (2022, p xv). Fraser labels this societal configuration 'cannibal capitalism' to indicate how the economic devouring of society amounts to society's eating of itself. She describes how capitalism's 'bouts of gluttony' lead to crises such that 'boom and bust' cycles come to resemble fits of bulimia. We can recognize communicative capitalism in this destructive hunger, and see the rise and fall of NFTs as one of its fits.

Turning, once again, to Disaster Girl, the sale of the meme can be understood as part of a flurry to use the new opportunities offered by NFTs to assert ownership over hitherto freely circulating signs and, hence, to cash in on the value accrued through the signs' circulation (Kale, 2021). As such, it is part of the boom cycle of NFTs specifically and cryptocurrency more generally. Since the meme was sold (in April 2021, remember), the crypto and NFT bubbles have, as mentioned earlier, burst spectacularly, illustrating what happens when speculative imagination fails. Financialized affect organizes the flows of signs, establishing the value of, say, Disaster Girl, but also orchestrates societal hype cycles around broader phenomena like NFTs and cryptocurrencies.

While cannibal capitalism, as Fraser (2022) discusses it, might have proceeded to eat its own heart out even without digitalization, the two do seem to feed on each other in particular and particularly voracious ways. In her highly influential work, Shoshana Zuboff (2019) generalizes this point in a sweeping critique of what she terms 'surveillance capitalism'. On digital platforms it is not just that – as the saying goes, 'if the service is free, you are the product' (Lanchester, 2017). The level of exploitation cuts even deeper than that, as digital traces of human lives (aka data) become the raw material for a burgeoning industry of extraction, commodification and control (Zuboff, 2015). Again, we may recognize communicative capitalism in the extraction of value from digital surveillance.

In Zuboff's words, the reduction of human subjects to points in the flow of data leads to 'a death match over the politics of knowledge in our information civilization' (2022, p 5). Or, in Dean's analysis, when circulation becomes 'the message', the message is lost, as 'a contribution need not be understood; it need only be repeated, reproduced, forwarded' (2005, p 59). Thus, Dean argues:

> Circulation has eclipsed meaning. That something is shared online does not depend on what it means. It depends on its affective capacity: does the shared item manifest outrage: is it funny and diverting? We attend less to the meaning of an utterance than to its affective dimension, which is most powerful when it contains different, conflicting meanings. (2019, pp 331–332)

Again, I agree with Dean's view of current circulations, which lead to 'the materialization of ideals of inclusion and participation in information, entertainment, and communication technologies in ways that capture resistance and intensify global capitalism' (2009, p 2) – or what I term the closing of the rhetorical mind. However, I disagree that meaning can only be reintroduced from outside of the processes of affective circulation. Rather, I believe those processes to be the very definition of meaning formation – affective circulation, as I understand it, *is* reasoning with a motive. The problem, in my view, is not with the circulation of affect, but with its currently dominant modality. The problem is that reasoning and affect have become severed, and the solution is to (re)connect them – to facilitate meaningful circulation.

Unfolding this argument and attending to affective alternatives to current modes of circulation will be the main purpose of Chapter 6. For now, let us turn to the question of digital labour and to the issue of what it means to work under conditions of communicative capitalism. Here, I begin by attending to the ways in which people are becoming entangled in processes of 'prosumption', the portmanteau of production and consumption that indicates how we are, increasingly, producing *and* consuming digital content (Bruns, 2013). This both supports Dean's diagnosis of the depoliticization of communication under conditions of communicative capitalism and offers a first indication as to why repoliticizing communication cannot (and should not) involve ending affective circulation.

Playbour and the produsage of value in digital capitalism

Extrapolating from the specifics of communicative capitalism, we may define digital capitalism more broadly as 'the collection of processes, sites, and moments in which digital technology mediates the structural tendencies of capitalism' (Pace, 2018, p 262). Within this collection – or sociotechnical configuration – digital labour denotes the modes of mediation, the different ways in which value can be produced (or 'prodused', as Axel Bruns (2013) would say) at the intersections of technology and structure (Scholz, 2013). In the digital economy, such value production is often nominally free, informally organized and/or invisible to everyone involved. As such, it suggests opportunities for empowerment *and* exploitation, especially at the blurred boundaries between private selves and public performances, between leisurely and professional activities – between play and work. Before considering the agential consequences of the particular ways in which production and use get entangled in what might be labelled digital 'playbour' (playful labour and laborious play) (Kücklich, 2005), let's look closer at the characteristics of digital labour that facilitate this particular entanglement.

First, users of digital platforms perform free labour all the time: when they create, edit and/or share a meme, like someone's comment, take on the role of moderator for a group and so on. This is the type of labour that fuels financialized circulation and organizes digital spaces in and through affective flows. Further, this free labour often draws directly on users' offline experiences, turning them into online content – as when someone takes a picture of their child in front of a house on fire and that picture becomes the basis of a meme. And it affects users' lives (online and offline) in various ways – taking time away from paid work and other activities, shaping social relationships and causing emotional stress/gratification.

In terms of its emotional toll (and potential rewards), free digital labour closely resembles the unpaid work that is traditionally associated with caregiving and homemaking. Arlie Hochschild (1983) distinguishes between the emotion work one carries out in private life and the emotional labour one undertakes in professional settings. The latter denotes the 'extra' layer of labour associated with keeping customers happy, tempering conflicts between colleagues and otherwise making sure everything runs smoothly at work. Hochschild (1979) finds that both types of emotion management are highly gendered, with women carrying the heaviest load in private as well as professional settings. Although her study is set in the empirical context of the US in the 1970s, its conclusions are surprisingly – and depressingly – durable across space and time.

In digital contexts, the line between work and labour, the paid and the unpaid, is often blurred, as one constantly manages emotions across and between different platforms – just as this emotion management is often monetized, whether by the person performing it (for example, an influencer), the platform on which it is performed (the influencer's social media of choice) or a third party (the company having their product promoted by the influencer). Despite these shifts, the gendered and gendering patterns of emotion management persist across offline and online settings. Thus, users who are perceived as women (or as holding (differently) minoritized subject positions) continue to be expected to perform the emotional support functions of online communities, while being subjected to much more online harassment and hate speech than (cis-male, White, straight, middle-class, ablebodied) majority subjects (Menking and Erickson, 2015; Gardiner, 2018; Nadim and Fladmoe, 2021).

Moving from individual experiences to the social level of affective circulation, the rise of new populism can, as Karen Lee Ashcraft (2022) shows, be linked directly to aggrieved masculinity. Here, affect is mobilized in defence of traditional societal norms and hierarchies, doubling the affective labour of those struggling for recognition within the structures that their very existence supposedly threatens. Thus, individual experiences of gendered emotion management are coupled with social processes of (re)producing gendered norms.

Doing emotion management online may feel exhaustingly familiar to the many individuals who continue to be underpaid and overworked. Yet, a seemingly inexhaustible number of people (across different genders) are betting that their free digital labour may become profitable, as they turn to the growing informal digital economy. This second feature of digital labour, its informality, is typical of influencers and others who have carved careers out of doing what they're doing online. Further, it is associated with the gig work of digital platforms, from delivery and driving through handiwork and housekeeping to consulting and copyediting. Both types of informal digital labour have witnessed momentous growth. As of 2022, influencer marketing was a business worth $16.4 billion (Geyser, 2022), and in 2021 the global gig economy had grown to a stunning $355 billion, with expectations of more than doubling its size within the next few years (*Business Research Insights*, 2022). With such magnitudes, calls to regulate the newly founded industries can no longer be overheard, but the task of formalizing digital work remains incomplete (Kaine and Josserand, 2019).

Digital labour, whether formally or informally organized, sources people's lives for (corporate) profit, tearing at the already permeable boundaries of work and leisure, public and private selves. As Melissa Gregg (2011, 2018) demonstrates in her work on flexibility and productivity, the fluidity of digital work extends from the independent content creator through the precariously employed gig worker to the salaried professional who is increasingly expected to be intimate with and committed to their work – always 'on', always available and always ready to use their private selves to drive up company profits. While these tendencies are reshaping traditional workplaces and employee relationships, they are directly built into digital platform work. Thus, it is noteworthy that Uber drivers and Amazon workers are leading unionization efforts across the globe (Scheiber and Weise, 2022; Bansal, 2023), just as democratically owned platforms present themselves as emerging alternatives to the big business of Big Tech (Scholz, 2016). As traditional workers' rights are eroding, digital labourers are struggling to voice their demands and are finding ways of turning protest into practice.

Surely, one reason why gig workers are leading the fight for labour rights in the digital economy is that this group of precariously employed labourers have more to fight for than people whose employment is already covered by hard-won labour rights. Another reason is that the conditions of digital labour are visible in gig work. For many others, digital labour remains obscure, even when they are performing it. This last characteristic relates to the two previous ones in various ways; first, if the labour is free, it is sometimes because the labourer does not know they are working. They are the product or, indeed, the raw material of digital capitalism (Zuboff, 2019), which also implies that the actual labourer is not a human actor, but rather a technological agent that works on and with people's data. Again, often in ways that are fully invisible or only vaguely perceived by the users whose social media feeds – or work–life interfaces – are being algorithmically monitored and curated. Thus, it is rarely clear to users what exactly digital technologies do, let alone how they do it (Bucher, 2017).

Second, if the work is informal, is it really work? Or, rather, how can it become recognized as such? We may now see brightly uniformed delivery people and boldly painted taxi services all over the cityscape, but many forms of digital labour remain less visible because they are less formalized. While neighbours may know that (and when!) a flat is being rented out on Airbnb, this is not clear from the street. And regulation may now require labelling of sponsored posts, but influencers (and their customers) are always finding new ways of seeming 'organic', thriving on the invisibility of labour – or, rather, the indivisibility of labour and leisure.

Further, 'clickwork', a growing industry of creating and annotating data and performing other piecemeal or microtasks (for example, on Amazon Mechanical Turk) offers invisible support for digital infrastructures. The often very emotionally taxing work of, for instance, labelling training data for large language models or removing material that is illegal or illegitimate from these datasets is completely absent in presentations of the final product and made even more invisible by its typical placement in the Global South, far from the more affluent contexts of most end users (Perrigo, 2023). Similarly, 'click farms' are the hidden sources of a lot of the seemingly organic activity that drives online circulation, adding followers to accounts, listens to songs, likes to posts and so on. The invisibility of digital labour, then, can sometimes be ascribed to hushed and stealthy uses of digital technology, but, more fundamentally, it derives from the various ways in which digital labour does not announce itself as labour – or, indeed, seeks to hide the underlying relations of labour and capital.

A lot of invisible digital work – whether paid or unpaid, formal or informal, hidden or out in the open – is in no way playful, let alone enjoyable. Still, it is a central feature of the onlife experience that it not only mashes up production and usage, but also makes the distinction between work and leisure ever harder to maintain. Julian Kücklich (2005) coins the concept of 'playbour' for the duality of work that is play and play that is work. This can take the form of the unpaid labour that players of computer games perform to improve the games they play, as is Kücklich's focus. Playbour is also the potential to make money from gaming as a professional player, as a streamer, or as a merchant of 'skins' and other in-game items. It appears as the gamification of work (digital and otherwise). And beyond its particular enactments, playbour denotes a new mode of being oneself and simultaneously other, offline and online (Just, Storm and Bukuru, 2023). Whatever its particular form, playbour constitutes precarious subjects who pursue potential economic gain at heightened personal risk (Mejias, 2010), not just economically but also in terms of what it means – and may mean – to be a human subject.

You be my body for me

Playbour reconfigures labour relations, but also maintains one of their central characteristics, namely the labourer's active involvement in hiding capital's exploitation of them. Judith Butler and Catherine Malabou (2011) discuss the process of hiding exploitation in their conversation about the analogy of lordship and bondage in the work of Hegel. Here, Malabou says of

Butler that they ventriloquize 'Hegel by giving speech to the master: "the imperative to the bondsman consists in the following formulation: you be my body for me, but do not let me know that the body you are is my body"' (Butler and Malabou, 2011, pp 613–614). With this formulation, labour becomes the achievement of detachment; detachment between the one who does the work and the one who benefits from it, but also, and more fundamentally, detachment from the labouring body amounting to the subject's self-awareness as a body and a mind. As Butler says in their response to Malabou's initial formulation of the problem:

> I encounter myself at a spatial distance from myself, redoubled; I encounter, at the same time, and in the same figure, the limit to what I can call 'myself'. These two encounters happen simultaneously, but this does not mean that they are reconciled; on the contrary, they exist in a certain tension and this 'other' who appears to be me is at once me and not me. So what I have to live with is not just the fact that I have become two, but that I can be found at a distance from myself, and that what I find at that distance is also – and at once – not myself. (Butler and Malabou, 2011, p 625)

Although this may seem a hefty theoretical load to throw at digitalization, it is also a remarkably precise description of online experiences generally and digital labour more specifically: at once me and not me, fully immaterial, fully embodied – or, in Dean's account of the technological fetish, 'the technology acting in our stead actually enables us to remain politically passive. We don't have to assume political responsibility because, again, the technology is doing it for us' (2005, p 63). Dean emphasizes the human passivity that arises from technological activity, how not having 'a body to implement a strategy' turns 'all the suggestions in the world' into 'noise' (2019, p 338). I suggest that human *and* technological agencies emerge from the continuous ripping open and stitching together of their entanglements, that making visible the me-not-me relationships of digital labour is a first step towards its politicization.

At the same time as the body disappears in digital work, then, it re-emerges – or, rather, it is redoubled in various ways. Today, technologies may stand in for the 'bondsman' in the analogy, working for their human 'masters' while covering up the toil of their work. Yet throughout the process of digital labour, it appears that the technological body is indeed the human body. Whether manifested as the gendered voices of chatbots and other digital materialities (Phan, 2017), in actual human bodies, the uniform(ed) labourers of the gig and click economy who remain (almost) unseen as they leave food at our door or make sure our digital tools are suitable for work, or as humans working for machines, inversing the relations of power once more. As such, the digital body is and is not the human body; constantly confronting us with Butler's constitutive duality: 'I can be found at a distance from myself' (Butler and Malabou, 2011, p 625). I (the human subject, any human subject) am at once social and technological, working digital technologies and being worked by them, embodying digitalization and being embodied by it. Digitalization, in sum, marks a reworking of labour relations where the worker is repositioned as not a worker at all, but rather an independent agent, a freelancer, someone pursuing their own interests rather than those of capital, someone 'venturing' into their own 'game of life' (Neff, 2015).

However, with this configuration, the worker is neither freed from their body nor from the master's exploitation of it. Rather, the worker remains, as Fraser in her conversation with Rahel Jaeggi has so aptly put it, free to starve: 'they [workers] are unencumbered by the sort of resources or entitlements that could permit them to abstain from the labor market. Their freedom in the first sense [freedom to work] goes along with their vulnerability to compulsion inherent in the second sense [freedom to starve]' (Fraser and Jaeggi, 2018, pp 24–25). Or, as I imagine Dean might agree, under conditions of communicative capitalism, the worker is not free not to communicate.

As the value of digital labour derives from its circulation, it does not matter (much) what one says as long as it spreads. Yet, as Dean also seems to recognize, affective circulation, while currently churning to the tune of global capitalism, does hold the potential for alternative value creation since 'producing for capital and producing social relations happens through the same processes' (2019, p 335). The question of how to make communicative circulation meaningful, how to recuperate its political potential and what impedes such recuperation is central to the remainder of this book. For now, however, let's recap the argument so far, and revisit the question of what all this has to do with Disaster Girl.

Organizing digital communication

The rendering visible of digital labour and its various embodiments may seem to have taken us far away from the illustrative case of Disaster Girl, but the meme has, in fact, been circulating in the empirical undercurrent of the conceptual discussion all along. At the surface level, the selling of Disaster Girl – and the broader monetization of memes – is just one more way in which affective value equals economic value under current conditions of digital capitalism. However, at a deeper level, Disaster Girl *is* the precarious subject, split from her own body yet working on – and through – it, spurred on by numerous others, human and more-than-human actors, who both contribute to the circulation of the meme for free and seek ways of profiting from it.

As such, the meme illustrates a fundamental condition of possibility for organizing digital communication: when partaking in digital processes of affective circulation, individuals are invited to take up the combined role of the prosumer, which spans all aspects of personal experience and decreases the ability to critically reflect on one's own position, covering up the constitutive split of the subject while also making it apparent – and profiting from it. Such 'self-commodification' (Lair, Sullivan and Cheney, 2005) is productive of the neoliberal subject of communicative capitalism. A subject that is constituted by technologies of power (and of the self) as an enterprising individual (Houghton, 2019), always looking for opportunities to improve the product that it is – and for means of capitalizing such self-investment.

While the neoliberal subject buys into the 'idea of the market as the site of democratic aspirations, indeed, as the mechanism by which the will of the demos manifests itself' (Dean, 2009, p 22), its relations to others thereby become instrumental, defined by market forces rather than democratic principles and ideals. In its current configuration, digitalization organizes communicating subjects into neoliberal markets; unregulated spaces of unhindered acceleration in which everyone always gets exactly what they want when they want it and through unrestrained processes of incremental value maximization (Leys, 2001). Thereby, the neoliberal subject becomes wary of the influence of others, contributing to its own 'dedemocratizing' by de-emphasizing differences of opinion (Brown, 2015). In sum, communicative capitalism closes the rhetorical mind.

In the next chapter, the conceptual underpinnings and practical implications of this claim will be explored at the level of public debate, guided by the question of what it might mean to imagine and enact a democracy without a 'demos'; without the collective of citizens, 'the people', who is the etymological and normative basis of the type of government, which in so-called Western societies is assumed to be the only legitimate organization of a polity.

3

The Digital Transformation of the Public Sphere

The following is a personal anecdote, but it is a story that I have heard others repeat often enough to suggest it may resonate collectively: shortly after Elon Musk took over Twitter on 27 October 2022 (and well before he renamed the platform 'X' in July 2023), my account, which had hitherto led a very quiet life, began to attract followers of a different sort than the academic types I had previously connected with: followers with single names and no or minimal profile descriptions, followers that felt 'other' – in fact, other than human. It seemed I had entered the orbit of the Twitter bots, which were central to Musk's argument when he was trying to wiggle out of the takeover (Clayton, 2022), but whose actual numbers and natures remain contested (Posard, 2022; Roth, 2023a). At the same time, my feed, which used to reflect my academic interests and connections, began to feature different topics and tastes. For instance, posts by Elon Musk now featured prominently, even though I did not follow him – an experience that seems to reflect a general tweak of Twitter's algorithms in those early days of the Musk takeover (Roth, 2023b).

Not only did my feed change in terms of profiles and content; it also took on a different feel than I was used to. Twitter, it seemed, had become a more hostile place, full of posts that were either antagonizing (pitting their authors against the positions of other users) or downright 'cancelling' (claiming the illegitimacy of others). This may, at least in part, be due to my academic connections' migration to Mastodon and other alternative social media platforms as they joined ranks with the many other users who left Twitter at the time (Huang, 2022) and left my feed in want of activity. But it could also be that more hateful content was being posted and promoted on Twitter in its post-Musk, pre-X days, as several studies have, in fact, found to be the case (Frenkel and Conger, 2022; Miller, 2023). While I did not experience this development as a good 'match' with my personal tastes, it certainly exposed me to different types of content and opinions than I was used to. And it did so in ways that I did not fully understand, prompting me to develop my own 'folk theory' (Ytre-Arne and Moe, 2021) of why my previously nicely personalized 'academic Twitter' had become submerged in what seemed a cesspool of polarization.

Stepping back from this anecdote, we may ask what it indicates about the current state of public debate. Musk himself has claimed that the platform he bought is 'the de facto public town square', but this position is highly contested and/or sorely lamented (Pariser, 2022). Some suggest Musk has 'broken Twitter' (Pearson, 2023), whereas others see recent developments as fresh evidence of an older claim that Twitter is 'the worst kind of public sphere' (Lilleker, 2014; Duncan, 2022). As such, we may assess the development of 'the platform formerly known as Twitter' through the lens of a broader debate as to whether social media are 'breaking' or 'making' the public sphere (Fuchs, 2021; Foust and Pratt, 2021). In diving into this question, I shift the focus from processes of digital communicative organizing to their infrastructural conditions of possibility, seeking to offer a conceptual framing of the present state of digital democracy. In other words, this chapter deals with the issue of whether and how digital communication can organize (for) democracy. The controversy around Musk's Twitter takeover offers a glimpse of the broader stakes of this issue: the interrelations between the political economy of digital platforms and the democratic potentials of digital publics. In other words, is digital democracy, in fact, democratic?

This chapter sets the conceptual stage for addressing the question of whether and how digital transformations threaten and/or support democracy, which subsequent chapters (Chapters 5 and 6) will then detail for the present and near-future sociotechnical configurations of digital public debate. Here, I begin by positioning the public sphere in relation to democracy, then consider the implications of that position from the perspective of the public and its members, asking, once again, the age-old questions of what a public is and what it means to be constituted as a member of a public. Thus, I will establish the structure, organization and modes of address that constitute public engagement, theoretically speaking.

Here, I am inspired by Axel Bruns' (2023) recent suggestion that we should think of networks of publics that are formed at the intersections of the personal level (organized as modes of address or what Bruns calls flows), the level of issues and interests (the organization of interconnections) and that of public spherules or, indeed, spheres (organized as structures). Bruns advocates the integration of normative theories and empirical studies but is himself primarily committed to establishing 'an empirically founded model' in the form of a 'toolkit'. Seeking to contribute to the further integration of theoretical and empirical work, I begin from the opposite direction, asking how existing theories may inform empirical insights.

With this aim in mind, I turn to the discussion of the digital transformation of the public sphere, gauging it across the structural, organizational and participatory levels of public engagement. Thus, I will first examine the structural and organizational dimensions of digital publics, thereby diagnosing current developments from the sociotechnical perspective: how *do* digital technologies transform the public sphere? Next, I will seek to connect these changes to previously established notions of the public sphere in order to discuss their consequences for democratic legitimacy. Here, deliberation is, initially, assumed to be the gold standard to which practices of public debate must be held. Subsequently, however, I question this assumption as digital modes of address are explored. Thus, the perspective shifts from a top-down to a bottom-up conceptualization of the digital public sphere, asking whether and what alternative conceptual understandings might emerge from the articulatory practices of actual digital publics.

The chapter ends by considering whether and why it even makes sense to continue talking about publics, offering a rearticulation of Bruns' (2023) networks of publics as the starting point for analysis and assessment of the democratic potentials of processes of digital engagement. In sum, this chapter offers the conceptual basis for exploring the closing of the rhetorical mind as a societal problem – a problem of democratic legitimacy. As such, the chapter is less driven by empirical detail and more attentive to conceptual explanation, but I assure you it is not just another review of the existing literature on the digital public sphere.

Not another literature review

It has become commonplace to write the story of the digital public sphere as one of democratic optimism followed by disillusion (de Blasio and Sorice, 2020; Cohen and Fung, 2021). First, events of the early 2010s – notably, social movements like the Arab Spring and Occupy Wall Street – sparked hope that digital media might offer platforms for (progressive) social change (Mazzoleni, 2015; Kraidy and Krikorian, 2017). Then came the fall from grace with the election of Donald Trump as US President in 2016 and the ensuing unravelling of the Cambridge Analytica scandal, which alerted members of national and global publics alike to the extent to which their opinions can be manipulated in the digital realm (Heawood, 2018; Chambers, 2021).

However, these are not historical developments, not a rise and fall of the democratizing potential of digital technologies. Rather, digital potentials for advocating and destroying democracy coexist, as may be illustrated by the near-simultaneity of events like the Brazilian Congress attack on 8 January 2023 (emulating the Capitol Hill riots of 6 January 2021) and the uprising in Iran, which began in September 2022 and continued to thrive in spaces of digital organization long after it had stopped drawing headlines in traditional media. Both these events were facilitated by mobilization on digital media and shaped by their online perpetuation (Kelly and Piper, 2023) – and their synchronicity indicates the multifarious uses of digital technologies for opposing/promoting democracy.

As mentioned in Chapter 1 (p 13), Kranzberg's (1986) dictum about the contingency and non-neutrality of technologies can be paraphrased for today's digital democracy, stipulating that digital technologies are neither inherently detrimental for nor supportive of democracy, and nor can their application to processes of democratic opinion formation and decision making avoid shaping these processes. The question is not one of cause and effect (how does digitalization affect society?), but of sociotechnical relations (what sorts of action do digital affordances invite and how are those invitations taken up?). Shifting the focus in this way enables us to circumvent much of the recent literature on digital democracy (although see Lorenz-Spreen et al (2023) for an overview). Still, exploring the question of how democracy is done digitally requires definitions of the key concepts of publics and public spheres, which must be situated in relation to democracy and digitalization.

Publics are queer creatures

The idea(l) of the public has ancient precedents, to which I shall turn in Chapter 4, but for now it will be discussed as a distinctly modern phenomenon – that is, as conducive to the process of individualization and rationalization in and through which human subjects become aware of their own unique identities and seek to organize society accordingly (Giddens, 1991). Modern society, in this sense, is organized around the precarious relationship between individual subjects and collective institutions. It is, as Gilles Deleuze says in the preface to *Difference and Repetition* (originally published in French in 1968), organized around simulacra: 'modern thought is born of the failure of representation, of the loss of identities, and of the discovery of all the forces that act under the representation of the identical' (2004, p xvii). As I asked in Chapter 2, what happens to democracy when the demos does not exist, when the state does not have a body? This is a quintessentially modern question for which the modern answer has been to talk the people into being as a self-reflexive non-entity; to paraphrase Butler and Malabou's (2011) discussion, the public is the state's body that is always found at a distance from itself.

Here, the public sphere – or 'publicness' as is the clunkier but more accurate translation of the German 'Öffentlichkeit' – emerges as the mediator between individual citizens and the state apparatus from which democratic institutions derive their legitimacy. Publicness is the process of recognizing the self in the other and the other in the self (Habermas, 1998a) around which modern democratic societies are organized. Modern democracies, we might say, are based on a foundational paradox: the principle of representing and enacting a 'will of the people', which does not, in fact, exist. Hence, the question of how this principle may be translated into practice, how 'the people' can be established as a willing and able collective subject with general interests and common opinions, is key to current democratic theories (at least those in the normative vein). And upholding the simulacrum of representation remains the underlying aspiration of contemporary democratic societies – or, as Dean writes, borrowing from the work of Laurent Berlant, 'contemporary mediatized technoculture' configures publics as 'spectacular forms of identification' (Dean, 1999, p 160).

Since its publication in 1962 (and English translation in 1989), Jürgen Habermas' *The Structural Transformation of the Public Sphere* has been pivotal to explorations of the democratic potentials of private-public mediation. Habermas' inquiry into the emergence and transformation of 'a category of bourgeois society' (as per the subtitle of the book) has been much celebrated, heavily criticized and repeatedly reinterpreted. This is not the place to offer another (mis)understanding of the work in its entirety. Rather, I will quote Michael Warner, who reminds us that Habermas' is an 'immanent critique' in the tradition of the Frankfurt School, aiming 'to show that bourgeois society has always been structured by a set of ideals that were contradicted by its own organization and compromised by its own ideology' (2002, p 46). The ideal, here, is that of 'the sphere of private people come together as a public' by means of 'people's public use of their reason' (Habermas, 1989, p 27, quoted in Warner, 2002, p 48). And the contradictions and compromises refer to the ways in which the organization and modes of address of 'actually existing' public spheres have failed to realize the ideal's 'emancipatory potential' (Warner, 2002, p 46).

When Dean asserts that 'Habermas claims that the public sphere, a critical discussion potentially including everyone, is lost at precisely that moment when it could actually appear' (1999, p 163), this is an astute but not entirely precise observation. The public sphere, in Habermas' conception, was never fully realized – and hence could not be lost. What is lost in Habermas' account of the first structural transformation (and this is where Dean's observation rings true) is the rationality of public debate. At the very moment that everyone can participate, it turns out that inclusivity does not in itself realize the potential for emancipation (and enlightenment) embedded in ideals of rational discourse (Habermas, 1990) – or, rather, that people participate in ways that are not consistent with these ideals.

Still, Habermas insists on the unfulfilled democratic potentials of the public use of reason, implying that 'publicness' is a norm rather than a fact, meaning that it is an ongoing achievement rather than an accomplished one, a 'process' or 'network' rather than a 'sphere' or 'space'. As he puts it in one of the more recent considerations of how democratic institutions may negotiate their commitments to the often-contradictory forces of facts and norms:

> The public sphere can best be described as a network for communicating information and points of view (i.e., opinions expressing affirmative or negative attitudes); the streams of communication are, in the process, filtered and synthesized in such a way as to coalesce into bundles of topically specified *public* opinions. (Habermas, 1998b, p 360, emphasis in original)

Thus, the opinions of the public emerge in and through the process of forming the public, a duality that forms the basis of Habermas' (1994) procedural theory of deliberative democracy. The simultaneous process of establishing public opinions and shaping collective identities constitutes polities *and* represents them politically.

The legitimating potential of this deliberative process is most clearly spelled out in Habermas' work on the constitution of the European Union (EU), which in his opinion is a particularly appealing polity, precisely because its public is not given and because it cannot assume the loyalty of a 'people'. Instead, the European polity is constituted along with its public – in and through deliberative processes that lend legitimacy to the political institutions (constituting the EU) and identity to the polity (constituting the EU citizenry) (Habermas, 2001a).

Here, 'constitution' doubles as a denominator for the process of creating something (in the sense also applied in Chapter 2) and the name of the constitutional text of a polity (that is, a legal document). To Habermas, these two meanings merge in the process of constituting the EU: the polity and its people, in his interpretation, come into being in and through the process of developing the legal text, amounting to a sense of collective allegiance – and a form of political legitimacy – that is conceptualized as 'constitutional patriotism'.

With its focus on constitution (as process and product), Habermas' democratic theory is radically procedural: 'the people' do not identify with each other substantially, but in and through their common commitment to legal principles, and the polity derives its legitimacy from following those principles rather than from the resulting policies. Or, in Habermas' own words:

> A previous background consensus, constructed on the basis of cultural homogeneity and understood as a necessary catalysing condition for democracy, becomes superfluous to the extent that public, discursively structured processes of opinion- and will-formation make a reasonable

> political understanding possible, even among strangers. Thanks to its procedural properties, the democratic process has its own mechanisms for securing legitimacy; it can, when necessary, fill the gaps that open in social integration, and can respond to the changed cultural composition of a population by generating a common political culture. (2001b, pp 73–74)

All of this may sound rather abstract, verging on the impracticable, and, indeed, anyone who has studied the process of constitutional reform in the EU will tell you it is a far cry from how this (seemingly interminable) process actually plays out (see for example Martinico, 2022). In fact, this is something I have spent more time contemplating than I now care to think about; my earnest efforts to understand the EU's constitutional process all boiling down to the laconic conclusion that when the European political institutions invite citizens to public debate, nothing happens (Just, 2005). This is not to say that people do not engage in debate around European issues (or, indeed, in European debate); they just tend to get excited about other issues and debate them in different manners than those stipulated by Habermasian ideals of deliberative democracy and emulated by the institutional entrepreneurs of the European polity (Just, 2016).

Besides missing the mark of actual empirical processes of opinion formation (in today's European context), Habermas' position on the (unfulfilled) deliberative potential of the public sphere has been criticized for its normative implications of bracketing private interests and personal dispositions as a prerequisite for the public use of reason (Fraser, 1992). As feminist and postcolonial scholars have convincingly demonstrated, the assumption that participants in public debate can disregard their particular embodiments and speak from 'nowhere', as it were, leads to privileging certain types of bodies and speaking positions that are consistent with – and benefit from – currently dominant societal norms. Deliberative ideals perpetuate epistemic injustice and ontological exclusivity, they restrict rather than enhance democratic legitimacy (Allen, 2012; Dieleman, 2015; Banerjee, 2022).

Warner is keenly aware of this problem, attentive as he is to the emergence of counterpublics whose 'protocols of discourse and debate remain open to affective and expressive dimensions of language' and who 'make their embodiment and status at least partly relevant in a public way by their very participation' (2002, p 58). Yet, Warner insists, publics are not just defined in opposition to counterpublics, as the results of exclusionary practices, but are themselves much more porous than they tend to let on.

In arguing this case, Warner goes so far as to suggest that publics are 'queer creatures', not just reliant on an excluded 'other' but defined by their very being besides themselves. In other words, although a public can appear as a 'social totality', as equivalent to 'the people', it is, in fact, a *social imaginary* rather than a *social fact*: 'A public might be real and efficacious, but its reality lies in just this reflexivity by which an addressable object is conjured into being in order to enable the very discourse that gives it existence' (Warner, 2002, p 67). Publics, then, are reflexively self-constituting, or, in Warner's words, *self-organizing*; a public, any public, is a 'a space of discourse organized by nothing other than discourse itself' (2002, p 67). This means that, with a concept Warner borrows from Hannah Arendt, publics are 'world-making', creatively fashioning a common world (Warner, 2002, p 59). And it means that publics are always multiple, not one comprehensive sphere, but numerous and diverse processes of self-organizing, of coming into being as publics. Publics are always besides themselves.

Still, self-organized world making, however multiple and multifarious it may be, is never unrestrained. Rather, it is an inherently bounded process of negotiating norms, making visible that which is otherwise hidden from the public, exposing the contingency of that which is taken for granted. Of hiding and making visible the ways in which publics are different from people – and people are different from polities, turning democracies into illusions of unity.

Whereas Dean (1999, p 158) argues that publics are inherently organized for consumption and therefore hold no potential other than that of legitimating the status quo, Warner insists on their emancipatory potential, although he construes this potential in very different terms than those used by Habermas. Here, Warner's exposé of the constraints placed upon those who wish to fashion publics differently is worth quoting at length:

> Because this is the field that people want to transform, it is not possible to assume the habitus according to which rational-critical debate is a neutral, relatively disembodied procedure for addressing common concerns, while embodied life is assumed to be private, local or merely affective and expressive. The styles by which people assume public relevance are themselves contested. The ability to bracket one's embodiment and status is not simply what Habermas calls making public use of one's reason; it is a strategy of distinction, profoundly linked to education and to dominant forms of masculinity. (2002, p 51)

To Warner, then, norms of public reasoning delimit the emancipatory potentials of publics, and democratic legitimacy is not built by conforming to those norms but by opening up spaces for their contestation. What also (re)emerges, here, is the question of the body; or rather, of the subject that can perform the sovereign body of the public without revealing that 'the body you are is my body' (see Chapter 2, p 38). This question, as I will now seek to show, is one of in- and exclusion, of the subjects that prevalent public norms exclude and of how the norms can become more inclusive – or, as

Dean might have it, how inclusion can be practiced as a gain rather than a loss. It is a question of freedom.

Subject to freedom

Ultimately, what the modern concept of the public offers is a sense of freedom, but a freedom that is suspended between two different intrusive forces: on the one hand, that which it cannot express or accommodate – the private; and, on the other hand, that which it must guard against or delimit – the state. So, what is the subject position that emerges as the bearer of such public freedom? Who can be recognized as the sovereign democratic subject? Is this subject consumed by the neoliberal subject of digital capitalism that emerged at the end of Chapter 2 or does it somehow hold the potential to set that subject free?

Questions of subjectivation are at the centre of Michel Foucault's writing, which begins from the assumption that subjects emerge from relations of freedom and power:

> Power is exercised only over free subjects, and only insofar as they are free. By this we mean individual or collective subjects who are faced with a field of possibilities in which several ways of behaving, several reactions and diverse comportments, may be realized … Consequently, there is no face-to-face confrontation of power and freedom, which are mutually exclusive (freedom disappears everywhere power is exercised), but a much more complicated interplay. (Foucault, 1982, p 790)

While Foucault is interested in the regularities of subjectivation, the ways in which power shapes subjects, he locates these relations at the level of both the state (technologies of power) and the subject (technologies of the self) (1988, p 18). Thus, Foucault says, power is not in itself 'a renunciation of freedom, a transference of rights, the power of each and all delegated to a few … the relationship of power can be the result of a prior or permanent consent, but it is not by nature the manifestation of a consensus' (1982, p 788).

This is important in the present context because it points to the tensions between a public's norms, its dominant modes of articulation and their possible subversion, and it indicates the ways in which public opinions can remain (or become again) contested even after a majority position has been formed, a decision has been made or a law has been passed. Thus, while Dean (2005) posits that under conditions of digitalization democracy has become an irrecuperable means to capitalist ends, implying that the democratic subject no longer exists or is no longer free, Foucault promises that all norms of subjectivation, however solid they may seem, can always be reconfigured, for better or worse.

To exemplify, let us consider, briefly, the rise and fall of abortion rights in the US. This vexing process illustrates how an issue (women's reproductive rights) that was, historically, relegated to the private sphere and kept far out of public sight could be, first, put on the public agenda and, second, obtain legal protection by reference to the very right to privacy. Many early articulations of the right to abortion, and of women's rights broadly speaking, were patently 'uncivil', infringements on the public sphere by means of protest rather than 'public reason'. In particular, as abortion was illegal, speaking about one's own experiences of it amounted to confession of a criminal act. Thus, articulating personal experiences became a powerful vehicle for bringing a private matter to public attention in a way that made space for other reactions than the simple denouncement of an illegal practice. But it was a legal case, conducted in accordance with the procedures of the US Constitution, that in 1973 confirmed US citizens' right to abortion.

Somewhat ironically, this case, *Roe v Wade*, passed its judgment on the basis of the right to privacy, sending abortion back to the private sphere, but with reversed effects – suggesting women's right 'to choose' would be protected and should henceforth go uncontested. As we now know, this has turned out to be a more ephemeral state of affairs than one could wish for. And in 2022, when the US Supreme Court decided to overturn *Roe v Wade* and revoke the constitutional right to abortion, proponents of reproductive rights were called to re-activate older forms of protest as well as to seek out new modes of subversion.

The decision also led to broader concerns as to which privacy rights might be up for grabs next (Morse, 2022) as well as to a reconsideration of the reasoning that upheld the decision in the first place. Privacy, with its implications of *freedom from*, of protecting the individual subject from infringements from others, might not always be the strongest form of protection, as has indeed become clear with the repeal. As such, the right to abortion might possibly have been better protected if it had been framed as a *freedom to* – couched within the right to personal autonomy (Greenesmith, 2022).

The matter of reproductive rights as discussed here (in the context of US legislation) alerts us to the intricate relations between the private and the public, indicating, first, how concerns hitherto thought of as private can claim public attention. Thus, it illustrates shifts in the configuration of what can be a matter for public reasoning, who can articulate such matters and how they can be discussed. Second, it shows how the freedom of individual subjects may be protected *by* rather than *from* public intervention (here, also in the sense of legal action), but it also illuminates how rights, once attained, can be revoked, making clear that freedom is not an end state but a process, not an act of liberation but an ongoing practice. This is why, as Maggie Nelson (2021) explains, Arendt 'sneers' at the suggestion that freedom can be a private or individual experience: 'Without a politically guaranteed public realm, freedom lacks the worldly space to make its appearance' (Arendt, quoted in Nelson, 2021, p 13). We may read the establishment and subsequent overturning of reproductive rights as an instantiation of the expansion and reduction – the waxing and waning – of the 'worldly space' for exercising freedom.

This returns us to Wendy Brown's concern with neoliberal de-democratization (see Chapter 2, p 40), which can now be reconceptualized as a 'crisis of freedom' (Brown, 1995) in which individual feelings of 'empowerment' legitimate broader antidemocratic tendencies of:

> liberalism that prizes individual liberties and rights almost without bounds, whether it's the right to spurn health mandates, the right to buy whatever kind of object you want regardless of how it pillages the earth, or the right to say whatever you want regardless of how violent and how damaging it may be. (Brown, quoted in Khachaturian, 2022)

For Nelson, this amounts to succumbing to 'the economic ideologies that align freedom with the willingness to become a slave of capital' (2021, p 9). Thus, public debate becomes a vehicle of control or, in Dean's words, 'communicative exchanges, rather than being fundamental to democratic politics, are the basic elements of capitalist production' (2009, p 56). Here, the subject is reduced to an economic actor, stripped of the political agency to explore freedoms that lie beyond the realm of 'prosumption'.

Warner also raises this concern in positing the practices of self-organized publics in opposition to the order of capitalism:

> The powerlessness of the person in [a totalitarian] world haunts modern capitalism as well. Our lives are minutely administered and recorded, to a degree unprecedented in history; we navigate a world of corporate agents that do not respond or act as people do. Our personal capacities, such as credit, turn out on reflection to be expressions of corporate agency. Without a faith, justified or not, in self-organized publics, organically linked to our activity in their very existence, capable of being addressed and capable of action, we would be nothing but the peasants of capital – which, of course, we might be, and some of us more than others. (Warner, 2002, p 69)

To Warner, as to Arendt, participation in publics is what enables individuals to fashion worlds in which they can act as free subjects, but such world making is precarious and constantly threatened, not least by the forces of capitalism. Or, as Foucault puts it, 'liberation paves the way for new power relationships, which must be controlled by practices of freedom' (1994, pp 283–284).

If and when people are unable – or unwilling – to participate in what Habermas terms the 'streams of communication' that 'coalesce' into public opinions, being subject to freedom becomes a constraint on the individual rather than an opportunity for their self-expression. And when people are able – or, indeed, forced – to participate in those 'streams', the question is, as Dean suggests, whether their participation serves anyone or anything but the forces of capitalism. This question is particularly pressing under conditions of digitalization; therefore, let us consider the case for and against digital freedom.

Information wants to be free

The advent of digitalization affects conceptualizations of the free public in two distinct respects, shifting how we may think of free speech and free access. Musk's assumption that Twitter is the 'de facto public square' leads him to conclude that 'failing to adhere to free speech principles fundamentally undermines democracy' (2022). And Mark Zuckerberg's deadpan answer to the senator who, in a 2018 congressional hearing on the Cambridge Analytica scandal, asked how Facebook could remain free of charge suggests the extent to which such 'free speech principles' have, in the context of digitalization, become subject to market logics: 'Senator, we run ads' (Zuckerberg as quoted in, for instance, Burch (2018) and circulated widely across the internet in a plethora of memes and other running commentary). It may be that, as the saying goes, information wants to be free – and that digital media support free speech. But freedom has, in both respects, become thoroughly monetized.

José van Dijck, Thomas Poell and Martijn de Waal (2018) discuss the political and societal consequences of these technological and economic developments, using the umbrella term of 'the platform society' to denote a societal (re)organization where digital platforms are becoming integral to society and productive of social structures. More specifically, the platform society 'refers to a society in which social and economic traffic is increasingly channelled by an (overwhelmingly corporate) global online platform ecosystem that is driven by algorithms and fuelled by data' (van Dijck, Poell and de Waal, 2018, p 4). Here, the notion of platform is not just a technical one, but also, as Tarleton Gillespie (2010) points out, a political construct, suggesting opportunities for connection and participation while eliding issues of liability and responsibility. In other words, digital platforms, as operated by Big Tech companies, provide an infrastructure for interaction between users of all kinds – individuals, organizations and public institutions – that is 'geared toward the systematic collection, algorithmic processing, circulation, and monetization of user data' (van Dijck, Poell and de Waal, 2018, p 4). And while this organization has profound societal consequences, it enables platform operators to deny that they are responsible for the effects of digitalization.

Thus, scholars like Gillespie (2018) may find it blindingly obvious that 'all platforms moderate', but Big Tech companies nevertheless insist that they 'just' provide the platform, whereas everything else is at the hands of the users. However, users' hands are increasingly tied by digital platforms, tethered to the algorithmic processing of data. Hence, all interactions on digital platforms are shaped – moderated, we might say – by the ways in which the algorithms that organize the platforms process user data, the aim being to elicit user responses that serve the platforms' operations – be it a 'mere' expression of emotion in the form of an emoticon (love, angry, haha, wow and so on) or a more directed desire such as the sudden urge to buy leg warmers (see Chayka, 2022). The example given here may appear innocuous, but it indicates how 'liking' in the sense of interacting with online content has become an ubiquitous director of desire.

The algorithmic organization of user data is, today, a pervasive experience: digital platforms offer opportunities of connection and interaction, basically enabling users to take their businesses and pleasures online, from mundane practices of catching up with friends, watching a movie or buying train tickets to (potentially) transformative life events like finding a partner, applying for a mortgage or deciding who to vote for in the next election. Underlying what might be thought of as enhanced opportunities for control of private matters and participation in public discourse is a restrictive rationality that organizes activities and shapes user experiences, always beyond their power and sometimes without their knowledge.

Most notably, algorithms curate the information that users are offered based on comparisons between a specific data point (for example, a Google search) and the combined datascape of both the individual user and all other users (the patterns of your search history plus the choices of users who conducted searches like yours), meaning that what you previously liked and what other users 'like you' have shown a preference for shapes the options you will be given.

The accelerating hype cycles around AI potentially becoming sentient notwithstanding, 'there is no there there' (Margolin, 2016) (for more on this, see Chapter 6). Algorithms are not conscious actors, but procedures for pattern making and matching. They are decision-making procedures that operate by connecting specific data points to general patterns. Algorithms, in sum, are 'sets of automated instructions to transform input data into a desired output' (van Dijck, Poell and de Waal, 2018, p 9). The publics that emerge from these procedures are anything but self-organized.

Calculated publics

Adhering to a Foucauldian understanding of rationality as productive of systems and subjects alike, Claudia Aradau and Tobias Blanke (2022) argue that algorithmic operations amount to a particular mode of reasoning, a reasoning that appears to be neutral because it is inherently without motive. Thus, algorithmic reason 'makes possible governing practices and the production of datafied subjects through the promise of more precise knowledge and more efficient decision-making' (Aradau and Blanke, 2022, p 4). As procedures of knowing, algorithms change not only what we know but also the conditions of possibility for how we know what we know. Despite (or because of) its appearance, algorithmic reason is anything but neutral; rather, it forms new power–knowledge relations that are characterized by 'multiple, mundane, and messy operations deployed to conduct the conduct of individuals and populations, self and other' (Aradau and Blanke, 2022, p 5). Algorithmic reason operates on individuals as if they were populations, and on populations as if they were individuals. This is done through data (a point to which I will return shortly), as algorithms extrapolate directly from individual data points to collective patterns and back again, drawing conclusions based on pattern making and matching. Thus, algorithms make predictions based directly on individual behaviours, but only in so far as these behaviours relate to populational patterns (or, rather, to the patterns that algorithms have identified in their training data).

Whereas Warner (2002, pp 55–56) suggests that publics cannot be numbered or named, and Aradau and Blanke argue that such quantification and identification is at the heart of algorithmic reason:

> If the statistical government of populations focused on producing aggregates, categorizing risk groups, and assessing abnormalities, algorithmic governmentality is no longer concentrated on either individuals or populations, but on their relations. Beyond shared norms and normativities, algorithms challenge political projects of the common and emancipatory possibilities of action. Even when algorithms are thought to produce publics, these are often seen as de-democratizing subjects, a 'calculated public' as the network of subjects and objects linked together through the digital. (Aradau and Blanke, 2022, p 11)

In such calculated publics, algorithmic reason only becomes visible through its outcomes; 'the algorithmic presentation of publics back to themselves shape[s] a public's sense of itself' (Gillespie, 2014, p 168).

Individual users may be more or less happy with specific results, but 'algorithmic anxiety' (Jhaver, Karpfen and Antin, 2018) is a general byproduct of algorithmic reason, as it is usually impossible for individuals to know exactly how the outcomes they receive were calculated. For example, what factors are given what weight in determining which posts to feature in my social media feeds? And if a change occurred, such as an exposure to more vitriol, is that because of something I did or due to a recalibration of the algorithm? As indicated in the chapter's introductory anecdote, my 'folk theory' of what changed the Twitter/X feed is as good as yours. We, the users of digital media, may know that the 'calculated publics' we encounter are not 'social facts' 'independent of any discursive address or circulation' (Warner, 2002, p 71). And yet, we have few means of challenging the calculated circulation that is presented to us as fact-based. For instance, it is hard to ascertain whether a drop in engagement with one's content is the result of followers' waning interest or an algorithmically imposed 'shadow ban' (Savolainen, 2022). And, more generally, we have a hard time determining whether the content we see is 'fake' or 'real' (Nielsen and Graves, 2017). Besides contesting what we see (or don't see), we have few means of glimpsing, let alone encountering the publics that lie beyond the boundaries of our algorithmically curated feeds.

This creates tensions between what algorithms actually do and what users think they do (Bucher, 2017). For instance, 'echo chambers' and 'filter bubbles' have been found, empirically, to be much less pervasive than popular debates and conceptual discussions of the concepts would lead us to believe (Bruns, 2019). From the technological perspective, we may cherish this fact, but socially people still interact with algorithms based on their beliefs about what they produce. No matter whether echo chambers are real or not, the social imaginary of them can have real consequences – for example, leading to social disengagement even when the technological potential for engagement exists. Who knows if the social media algorithms have changed my feed, whether the changes are due to the behaviours of the people I follow (that is, their decreasing activity on Twitter/X after Musk's takeover) or, perhaps, to the proliferation of automated profiles? Whichever way I – and we – answer these questions, the trend will be the same: personalized feeds (whether we are happy with the personalization or not) become increasingly privatized, removed from 'the public use of reason'. In sum, algorithmic reason organizes calculated publics, which are not conducive to public self-organization, but, instead, turn individuals and collectives alike into data points for further calculations.

Datafied subjects

Algorithmic reason relies on data – and turns everything into data, thus supporting the process of datafication, defined as 'the quantification of human life through digital information' (Mejias and Couldry, 2019, p 1).

In the context of digital platforms, what is usually captured and made available to algorithmic reason is information about users' identities and behaviours that can be used to predict their preferences in terms of content to be exposed to and recommendations to be offered (van Dijck, Poell and de Waal, 2018, p 33). It is this information that is extracted, repackaged and sold to advertisers.

However, datafication is not just the basis of the dominant business model of digital platforms; this mode of value extraction also involves the formation of new data subjects. Nick Couldry and Ulises A Mejias (2019) discuss the process of datafied subjectivation in terms of data colonialism, arguing that current digital practices are reminiscent of past actions of colonial states *and* enact (neo)colonial relationships that are (re)productive of structural inequalities between digital platforms and data subjects. Just as digital circulation is literally financialized (see Chapter 2), so the human subjects who participate in such circulation are, Couldry and Mejias argue, literally colonized:

> In the hollowed out social world of data colonialism, data practices invade the space of the self by making tracking a permanent feature of life, expanding and deepening the basis on which human beings can exploit each other. The bare reality of the self as a self comes to be at stake … This reality, which each subject can recognize in each other, cannot be traded away without endangering the basic conditions of human autonomy. (2019, pp 344–345)

The datafied subject, then, is patently not free, but exploited in at least two ways: as data and by data. As it happens, these two modes of exploitation converged in the aftermath of the overturning of *Roe v Wade*, when the intimate data collection of period tracking apps became a matter of urgent concern (Saikia and Doshi, 2022). Faced with the threat that data trails might reveal practices of reproductive health that would now be illegal, users advised each other to delete their period tracking apps, but beyond such defensive action had no means of reclaiming the right to their own bodies, let alone their data.

Van Dijck, Poell and de Waal (2018) posit two other mechanisms of digital platforms alongside datafication: commodification and selection. Selection, I believe, is the thread that runs through algorithmic reason and datafication to produce commodification. Here, commodification is defined in similar terms to those used in Chapter 2 to describe the simultaneously empowering and disempowering experience of being able to monetize one's passion projects while often ending up working for free. Beyond the experience of the prosumer, it denotes the degree to which the self is commodified – wittingly as content and unwittingly as data.

In the platform society, all data is a (potential) commodity, a raw material that can be augmented by labour – and usually the companies behind the platforms pocket the resulting surplus value. This happens through processes of selection, of curating content to users and users to advertisers (and other buyers of user data) and of circulating datafied information among users as well as to third parties (van Dijck, Poell and de Waal, 2018, pp 40–41). All of this contributes to the calculation of publics, to their construction as social facts, organized externally by algorithmic reason rather than self-organized through discourse (cf. Warner, 2002, p 70). Thereby, individuals become increasingly isolated from the publics in which they partake.

While this claim may seem paradoxical, the point is that digital participation, whatever else it might be, is an alienating process of talking to an aggregate of others as well as of oneself. Even when experiencing digital interaction as participatory, maybe even emancipatory, we are simultaneously being counted and sorted, packaged as commodified versions of ourselves, which are sold on to others – and back to us. To paraphrase Butler once more, what I find at the digital distance from myself is at once me and not me.

Consequently, in the platform society, information may be free, but everything else is for sale – and the price is often paid in the currency of freedom. This means that individual subjects become more bound and opportunities for public participation less open. As processes of selection run through algorithmic reason, datafication and commodification, calculated publics are organized to favour – and in favour of – the profit maximization of platforms. Ultimately, the question becomes whether we can fashion publics when we are mostly talking to algorithmic curations of ourselves and, hence, whether Twitter/X and other social media are – or can become – arenas for controversial encounters. While I will explore this question in chapters to come, it is now time to settle the conceptual matter of whether digitalization may be reconcilable with deliberation.

Is Habermas on Twitter?

This question, as posed by Axel Bruns and Tim Highfield (2018), may seem purely rhetorical. And, empirically speaking, the answer is no, of course not. There are several Habermas profiles on Twitter/X, at least two of which are active and, somewhat surprisingly, seem to prefer to tweet in Spanish. But the 'real' Habermas does not stand up for social media. Conceptually, however, the answer is a bit more complicated, as it returns us to the issue of not just whether Twitter/X and other digital platforms are capable of forming publics, but how digitalization transforms the public sphere.

Leading up to the 60th anniversary of Habermas' work on *The Structural Transformation of the Public Sphere* (originally published in German in 1962), a number of publications have dealt with the issue of further structural transformations. For instance, Tiago Santos, Jorge Louçã and Helder Coelho (2019) consider how digitalization changes the agenda-setting capacities of traditional (news) media, showing empirically (through the case of the Brexit referendum) how the media agenda and the public agenda have become more dispersed and polarized. Offering a conceptual take on these developments, Evan Stewart and Douglas Hartmann (2020) consider the effects of 'networked individualism' in horizontally proliferating publics that can be directly accessed by their members. These authors identify three areas of structural transformation: first, new inequalities emerge around the 'digital divide' of access to digital technologies and knowledge of digital cultures; second, the many multiple publics are not connected to each other in a larger network, but are increasingly detached, adding further impetus to polarization; third, the issues of inequality and separation result from 'the structural traits of digital communication itself', overlaying the horizontal formation of publics with 'immense "vertical" power' (Stewart and Hartmann, 2020, p 176).

In the introduction to a special issue of *Theory, Culture & Society* that examines the ongoing structural transformation of the public sphere from different angles, Martin Seeliger and Sebastian Sevignani (2022) diagnose the consequences of digitalization along similar lines, arguing that digital affordances of direct access and unhindered interaction (circumscribing traditional media as agenda setters and gatekeepers) may hold democratic potential, but are, in their current enactment, democratically detrimental. The participatory potentials of these affordances are, they argue, hampered by dominant platforms' surveillance-based business models:

> Monitoring and evaluation of digital communication serves a cybernetic logic of control where communicative action becomes more effectively controllable, or even enables feedback-logical propaganda in the interest of those who have the data or know how to use them. This sets the course either toward a shutdown of deliberation in the sense of a mere networking of private opinions … or an increasingly acclamatory form of the public sphere, as expressed, for example, in simple 'like' or 'dislike' expressions, within a mode of 'privatized representation'. (Seeliger and Sevignani, 2022, p 12)

While digital media might enable open and inclusive communication, these affordances are, so the argument goes, warped by structural features, leading to fragmentation and polarization. From the perspective of democratic theory, what might have been desirable pluralization becomes problematic disintegration.

Habermas (2022a, 2022b) has added his own voice to this debate, expanding his reflections on the special issue in a book instead of condensing them in a tweet. In Habermas' diagnosis, the digitalization of the public sphere further erodes the correspondence between citizens, on the one hand, and the state, on the other hand, and therefore is detrimental to deliberative democracy of the type he favours. He points to two interrelated features as especially problematic: the platforms' rejection of editorial responsibilities, including their role as moderators, and the blurring of the distinction between public and private interests and subject positions. In this diagnosis, the empowering potential for participation that is inherent to digital platforms becomes democratically problematic because users are invited to participate in ways that do not live up to the standards of public reason, most notably 'the generalization of interests', and platforms do nothing to hinder the resulting circulation of falsehoods and defamations. This leads, Habermas fears, to the 'rejection of dissonant and the inclusion of consonant voices into their own limited, identity-preserving horizon of supposed, yet professionally unfiltered, "knowledge"' (2022a, p 166) – that is, to the combined dynamics of polarization and personalization.

The result, as Markus Patberg (2023) explains in a review of Habermas' book-length argument:

> is the emergence of semi-publics, whose members no longer regard the general public sphere as the place for the discursive clarification of validity claims, but see it as a realm of hypocrisy whose protagonists ignore 'the truth', i.e. what appears as such from within the self-referential spaces. What was once an inclusive space, integrating all citizens, is thus degraded, in the perception of some members of society, to just another sectarian semi-public.

The digitalized public sphere fails to support democracy, as citizens can neither see their views reflected in political decisions nor accept those decisions as the result of the public use of reason.

Given that such correspondence and acceptance are the constitutive principles of Habermas' theory of deliberative democracy, the current situation, in his view, amounts to nothing short of a democratic legitimacy crisis – or, as Dean (2003) writes, in her discussion of 'why the net is not a public sphere', which is simultaneously critical of the Habermasian ideal and the digital Real (see Chapter 2, p 26):

> [the Web] provides an all-encompassing space in which social antagonism is simultaneously expressed and obliterated. It is a global space in which one can recognize oneself as connected to everyone else, as linked to everything that matters. At the same time, it is a space of conflicting networks and networks of conflict so deep and fundamental that even to speak of consensus or convergence seems

> an act of naïveté at best, violence at worst. Both these dimensions of convergence and conflict hold without canceling each other out or resolving into a process of legitimation or some sort of will-formation that carries with it a supposition of rationality. (Dean, 2003, p 106)

As is apparent, Dean describes the situation in terms that are quite similar to the Habermasian critique of personalization and polarization. And, indeed, Dean has said of Habermas' earlier work that 'the problem is that his description is right, but his diagnosis is wrong' (1999, p 162).

I agree and align myself with Dean's alternative diagnosis 'of networked communication as lacking public sphere norms on the one hand, and as plagued by a surfeit of these norms, on the other' (2003, p 98) – that is, 'the ideal of the public works simultaneously to encode democratic practice and market global technoculture' (2003, p 102). Democracy, as practiced in and through the norms of public reasoning, becomes good for nothing but upholding capitalism.

Again, I side with Dean, but as will be substantiated in Chapter 4, I disagree with both her and Habermas' preferred alternatives. What democracy needs now is neither more deliberative consensus seeking (Habermas, 1990) nor enhanced partisan struggles for hegemony (Dean, 2003, p 110), but the re-appraisal of disagreement. What we need is not a return to old forms of public reasoning, whether in the bourgeois or the Marxist distillation, but the development of new norms of interaction. Some theories of the public sphere do, as the next section explores, support this position, seeking to change the norms of public reasoning from within processes of digital public debate and, hence, to salvage their self-organizing, democratic potential.

Only connect

Theoretically speaking, whether from the structural perspective of the platform society or the normative perspective of deliberative democracy, it is reasonable to say that Twitter/X and other digital platforms do not form public spheres. In practice, however, this claim makes absolutely no sense. If we understand publics as self-organizing around modes of articulation that creatively fashion publics, then digital publics of all sorts quite evidently self-organize all the time, the forces of calculation and datafication notwithstanding. As even the likes of Habermas and Dean recognize, digital platforms offer ample opportunities for participation and people do take up these opportunities with passion and vigour. It's just that the invitations to participate and the uptakes of them are a far cry from the norms of the public use of reason that Habermas espouses and from the political antagonism that is Dean's ideal.

Focusing on how users enact the invitations of digital affordances and how they actually communicate within the confines of algorithmically organized processes, it becomes clear that people have the potential to do many things online. Further, at least some of the things that people actually do are potentially democratic – for example, debating political topics and mobilizing around societal issues. In particular, if one zooms in on the ways that digital platforms support grassroots activism and social movements, it becomes possible to reassert the democratic potentials of digitalization, but it is also evident that this requires a reconsideration of the normative basis of democratic theory (Kaun and Uldam, 2018). Most notably, we will have to replace the ideal of deliberation, and while many alternatives present themselves connection is emerging as a particularly promising contender.

Dean (2005, 2009) views the connectivity of online social movements as part of the problem, arguing that it pulls in the direction of postpolitics. Still, the democratic potential of connection can be advocated along at least two lines. First, it has been argued that, in the context of social movements, 'connective action' is replacing 'collective action' as the dominant mode of mobilization. Connective action, as conceptualized by Lance Bennett and Alexandra Segerberg (2013), is a personalized form of activism that emerges from two key affordances of digital platforms: their invitation to open and inclusive participation. In other words, as users circulate their personal opinions and experiences, self-organized digital networks may turn individual positions into collective demands. This perspective indicates how common positions may arise even as individuals remain dispersed, suggesting that networks can, indeed, be made up of very loose ties and still form publics in the sense of collections of people that voice common demands. Here, personalization can lead to mobilization, creating what might be labelled collectives of individuals.

Second, the agential potentials of connection form the basis of a generalized theory of 'connective democracy', which suggests that the divisive and polarizing tendencies that currently dominate (digital) public debate may be ameliorated, even reversed, if people 'only connect' (Stroud, 2021). More specifically, this notion of connection facilitates the bridging of diverse groups who come together to discuss their differences and gain increased understandings of each other while not (necessarily) becoming more alike.

This may seem as idealized an understanding of democratic processes as that offered by deliberation, and, it should be noticed, connective action as defined by Bennett and Segerberg holds none of the normativity of connective democracy. Rather, connective action is an analytical concept, based on studies of actual online mobilization. Attuned to current sociotechnical conditions of possibility, analyses of connective action might, in fact, support Dean's view that online activism must succumb to the ideology of communicative capitalism, thus concluding, once again, that change from within is impossible. Conversely, connective democracy is more of a conceptual vision than an observed phenomenon – a theory of change rather than a practice of it.

Still, proponents of connective democracy hold that their vision is attuned to the affordances of digital platforms, offering a theoretical extrapolation of the invitations to participate and engage that are inherent to digitalization – and, importantly, proposing a more amenable evaluation of them. As such, connective democracy posits itself as 'deliberative democracy's grittier, more down-to-earth cousin' (Masullo and Overgaard, 2021, np) that begins from the fact of actually existing digital public debate and takes things from there. This, then, is a theory that suggests we look beyond the order of calculated publics to the beautiful mess that people make of them.

No manners

Connective democracy's insistence that we can still 'have nice things' on digital platforms is, I believe, a useful supplement to more dismal structural accounts. However, to see the amicable potential of digital interaction, one must downplay (if not outright ignore) the fact that users' digital engagements tend to be anything but 'nice' (Phillips, 2015). Or, as Zizi Papacharissi points out, maybe one should look elsewhere: 'it is … possible that our quest for civic behaviours has not produced the desired results because we have not been looking at places that civic behaviours now inhabit: spaces that are friendlier to the development of contemporary civic behaviours' (2010, p 78).

Even so, it is commonly noted in empirical studies of digital vernaculars that the friendliest of digital contexts are rude and unruly by deliberative standards; at their most agreeable, digital interlocutors tend to be inattentive to decorum or, perhaps more accurately, espouse a decorum of having no manners (Goode, McCullough and O'Hare, 2011; Chen, 2017; Richter, 2021). What, then, should we make of digital publics that self-organize around transgressive norms – for example, of satire and ridicule (Caron, 2021) – and scorn dialogue, reason and other established norms of public debate, whether in the vein of deliberative, connective or any other theory of democracy? Rather than leading back to a rejection of the democratic potentials of digital platforms, I suggest this query returns us to the driving force of affect (as introduced in Chapter 2 and as will be further developed in Chapter 7). Also, it leads us towards an ideal of interaction that is more amenable to contestation than connection (as will be developed in Chapter 4) and that sees democratic potential in networked publics (as I will develop here).

Digital publics, as Papacharissi (2014) proposes, are affective, open not just to expressions of feeling, but also to processes of meaning formation in which people come to terms with what they think based on how they feel. Hence, the affective and expressive dimensions of public debate are not, as proponents of deliberation tend to imply, necessarily barriers to common meaning formation; letting affect into the arena of public debate does not relegate people to each their own emotional rabbit hole, but is, instead, an inevitable and necessary feature of processes of meaning formation. Indeed, reasoning depends on its motives, and failing to recognize that rationality is inherently affective is, in my opinion (see Just, 2016), the biggest mistake Habermas has made. The circulation of affect may intensify feelings in divergent, even polarizing directions, but that is an outcome of particular processes of circulation rather than an inherent quality of their affective nature. Whichever way feelings flow, they are inherent to public debate, to be accepted and studied as a condition of possibility, not scorned or celebrated *a priori*.

Further, digital publics are issue-oriented, as Noortje Marres (2021) demonstrates, organized around topics that people care about and articulated as reasons that motivate them, rather than in terms of a generalized idea of 'the public interest', which may remain a political ideal (and a politically expedient steering mechanism), but is perpetually deferred, abstract and elusive. 'Indeed', as Dean argues, 'a democratic theory built around the notion of issue networks could avoid the fantasy of unity that has rendered the publicity in technoculture so profoundly depoliticizing' (2003, p 111). Within the theory of issue publics, it is recognized that the articulations of topics of concern are shaped by the media in which they are discussed, and that these discussions, in turn, shape the contexts of their own mediation (Marres and Moats, 2015). The formation of issue publics may be partial (not complete *and* with certain inclinations), but at least it is real – and holds real potentials for democratic debate. And it should be studied in and as specific sociotechnical constellations of issues, (human) actors and (digital) media.

The upshot is that when studied 'on the ground', in and through their modes of articulation, digital publics are multifarious and dispersed, sometimes fleeting. Further, digital publics adhere to other norms than those having to do exclusively with rational debates (although some publics may hold to these norms). Also, it is increasingly difficult to imagine *the* public as a unified and encompassing entity that could somehow legitimate political systems generally and political action more specifically. But that does not mean digital publics are necessarily undemocratic or that digital public spheres have no legitimating potential; all it means is that we must look elsewhere and look differently than theories of modern democracy have tended to invite us to do. Or, as Dean concludes her consideration of the democratic potential of issue publics, 'democracy … may well be a secondary quality that emerges as an effect or a result of other practices, but that can never be achieved when aimed at directly' (2003, p 111).

In the following chapter, I develop my alternative to the modern conceptualization of the public sphere, based on the classical notion of *controversia* – reasoning that welcomes grittier and more unruly modalities than can be accepted within the norms of deliberation. However, before digging into this work of conceptual recovery, it might be worth considering whether and why we should take the notion of publics with us.

What do we need the concept of publics for?

Given the theoretical and normative baggage of the concept of the public, it could make sense to simply abandon it, as Dean argues. Instead, we could, as Papacharissi (2010) suggests, go looking for participation in different places and through alternative modes, assessing what people actually do online by other standards than those offered by the norm of public reason. Thus, we might explore how 'democratic practices and modes of affiliation [could] be uncoupled from a notion of the public sphere and understood within a different political architecture' (Dean, 2003, p 111). Yet, I maintain that the concept of 'the public' offers the best starting point for exploring the democratic potentials of controversial encounters.

To be sure, reharnessing the concept for this task requires some conceptual manoeuvring. For starters, I recommend replacing the singular with the plural; the public only ever appears in and as publics. Even so, the concept of publics may need further reconsideration, but that is an effort worth making: insisting that digital publics *can* be democratic publics provides a necessary starting point for identifying the democratic potential of digital public debate and considering how it might be realized – not as one coherent public sphere, but as networks of publics and, importantly, counterpublics that are explicitly self-organized against the grain of algorithmic reasoning with its organization of datafied subjects into calculated publics.

The word 'publics' may not be for everyone, but it is, I believe, useful for the task before us. In particular, we may use the notions of issue publics and affective publics as starting points for the analysis and assessment of the democratic potentials of actually existing – self-organized and multifariously articulated – networks of publics.

Bringing the conceptual and empirical lines of argument together, let us note that the organization of individuals in and through communicative processes has never been as reasonable as the ideal of deliberative democracy requires (Dahlgren, 2005). Nevertheless, it is important to recognize that digitalization multiplies and disperses public spheres while exasperating norms of deliberation; if one conceptualizes the public sphere in Habermasian terms, then digitalization does, indeed, spell trouble. And I agree with some of the issues, most notably around polarization and personalization, just as I think we are worse off with algorithmic reason than with deliberation. However, it is also important to recognize, as Dean so forcefully argues, that the norm of public reason is complicit in the reproduction of current mores. Agreeing with both Habermas and Dean on these points, I nevertheless insist on seeing the potentials of as well as the trouble with digital publics.

For one thing, publics are, as Warner (2002) argues, discursively self-organized, existing in and as nothing but the communicative process itself. Any digital public worth the name is self-organized, leveraging the participatory potential of digitalization against the very tendencies of the algorithmic processing of data to organize 'calculated publics', which are, in fact, not publics at all. Importantly, the participatory *and* defiant character of publics, their self-organizing capacity to work with the affordances of digitalization and against the infrastructures of platformization, makes digital publics as susceptible to change as to stability.

Further, publics are organized around particular topics of concern rather than general interests. They are not (purely) rational but are organized affectively, even when affective intensification clings to – and reproduces – current norms of the public use of reason (Just, 2016), even when algorithmic reason structures affective circulation. While none of this may be ideal, it is real. And, hence, it forms a more adequate starting point for understanding how publics self-organize and, by implication, a more precise basis for evaluating – and perhaps improving – the currently dominant organization of digital publics.

Finally, digital publics are, indeed, algorithmically ordered and produce data subjects, thereby narrowing individual and public freedoms. In other words, digital platforms do invite calculated publics that throw individual users back upon themselves rather than into public discourse. However, these conditions, too, can be resisted and turned into occasions for taking liberties, as practices of algorithmic activism and data justice emerge, offering new modes of resistance and new opportunities for subject formation (Dencik et al, 2019, 2022; Bonini and Treré, 2024).

Publics, in sum, are sociotechnical configurations of a particular kind, whose potentials – democratic or otherwise – are only enacted in and through their organization. At present, the most readily available invitations to action may tend towards the organization of individual publics rather than public individuals, as the algorithmic production of publics invites ordering and numbering of individuals and impedes the 'creative fashioning of common worlds'. Hence, it may be that Twitter/X and other social media are not public spheres, but that is not to say that digital publics hold no democratic potential. The digital transformation of the public sphere is a contingent process, not a social fact.

The question is: how can self-organizing publics thrive within the confines of calculated publics? And the answer might be that they can't, that maybe we simply need to get off digital platforms – or at least off the commercially organized ones that only invite public debate as an excuse for harvesting our data. Maybe Twitter is already dead – and not just in order to be reborn as

X (David, 2023). And while Elon Musk may have killed Twitter, it is worth asking, as Sarah Jeong (2023) does, whether it ever had any power over its users' hearts and minds:

> In an alternate universe where Twitter had not been acquired by Elon Musk, where Twitter is not crumbling, do I think and feel differently? Do I have different opinions, do I support different causes? Is it at all possible that the firehose of the thousands and thousands of cheap and easy words changed my mind, altered the very arrangement of atoms in this world, solid mass on solid mass?

The death of Twitter, then, may not in itself alter the digital transformation of the public sphere, but it does illustrate the contingency of the process. As Jeong (2023) concludes, 'whether or not Twitter itself changed the world, its decline and imminent death will, at the very least, change our experience of how the world changes'.

Just as newspapers were once threatened by television and television lost out to social media, so social media can come and go – individually (just ask MySpace) or en bloc. However, no matter what the next transformation will look like, digitalization remains ubiquitous at present, shaping the empirical reality of public debate into networks of publics that build connections from the individual upwards, organizing personal publics, issues publics and publics of publics (Bruns, 2023).

Following Bruns (2023), this is the empirical reality we should tune into, judging its particularities rather than its distance to an abstract conceptual norm. And that is what I will do in Chapters 5 and 6. Still, even specifics can only be judged in relation to something – and that 'something', for me, is the potential for controversial encounter, to which I now turn. Thus, Chapter 4 establishes the historical roots and contemporary standards to which I will subsequently hold dominant and emerging configurations of digital public debate. In combination, the next three chapters will detail the organization of what I term controversial encounters of the first, second and third kinds; the first being a somewhat romanticized ideal that has never existed in empirical reality but may nonetheless help us critically assess controversial encounters of the second and third kinds.

4

Controversial Encounters of the First Kind: The Theory and Practice of Controversy

Modern publics are, as Michael Warner (2002) emphasizes, reading publics; their ideal typical communicative situation is one of individual encounters with public arguments in writing. The reader and the writer are set apart from each other and only meet in and through the text. Here, as Warner puts it, 'the most private, inward, intimate act of reading can be converted by the category of the public into a form of stranger relationality' (2002, p 84). This is not so for the publics of earlier iterations of democratic society. In ancient Greece and classical Rome, publics assembled *in* public to deliberate orally on matters of public concern and arrive at a decision on the spot. The relationality involved here is not one between strangers who reason privately about public matters, but between fellow citizens who debate their private interests publicly. Hence, in the classical conception, the public use of reason was not severed from the articulation of personal motives; rather than aiming for consensus, classical publics were organized around the principle of persuasion.

In this chapter, I turn to the common origins of Western democracy and rhetoric to excavate the classical concept of *controversia*, which I posit as an alternative to current ideals of democratic public debate. In so doing, I will situate the classical concept between modern theories of deliberative democracy and agonistic pluralism. As I go back to classical Greek and Latin theories of persuasion, I will draw on modern exponents of a controversial view of rhetoric, but I will also read this (classical and modern) literature through postcolonial and feminist critiques that expose the racist and misogynist underbelly of rhetorical thinking in 'the European tradition'. As such, I am attentive to the constraints of rebottling old concepts for new purposes, and rather than conducting a full conceptual history of *controversia* or a historical analysis of the development of the empirical genre(s) of controversy, I will recapitulate the clashes (controversies!) between different models of persuasion. In recuperating *controversia*, then, I seek to establish an ideal for public debate that is attentive to its own limitations and privileges the maintenance of difference over its various historical, structural and practical eradications (whether by coercion or persuasion). Thus, I develop the concept of controversial encounters to denote a form of public debate in which participants are able to recognize differences and sustain disagreements, meeting each other in the space that exists between consensus and conflict. In the present chapter, I establish an ideal typical configuration of this space, labelling it 'controversial encounters of the first kind'. This is an abstraction (*not* a historical reality) against which current configurations of public debate can be measured. To set the scene for all this, I first invite you to join me on an extended excursion to ancient Athens.

The Mytilene debate: a paradigm for deliberative rhetoric

Thucydides tells the story of how in 427 BC the Athenian assembly had to decide on what punishment to mete out to Mytilene, the main city of Lesbos, for having violated its status as a privileged ally of the Athenian league through armed revolt. After the first day of heated discussions, the vote resulted in the harshest possible punishment: killing all men and making slaves of the women and children of Mytilene. While the Athenians had initially been inflamed by the account of Mytilene's villainous treason, they subsequently became concerned with their own virtuous reason – and decided to consider the matter once more: 'The very next day', writes Thucydides (36.4), 'there was an immediate change of mind, and they began to reconsider the savage and extreme decision to destroy an entire city rather than just those directly responsible.'

Therefore, a new assembly was called and the matter was debated once more. Thucydides includes (or, rather, makes up) the full speeches of this debate, representing both positions equally (though not, it should be said, with equal loyalty). Arguing in favour of upholding the decision, the first speaker, Cleon, concludes:

> We must not, therefore, hold out to them any hope, whether secured by oratory or bought with bribes, that they will be pardoned just on grounds of human fallibility. They did us harm not through some involuntary deed but in a deliberate act of conspiracy, and one only pardons what is involuntary. (Thucydides, 40.0)

Diodotus, who represents the position that the punishment should be reduced, answers directly to Cleon's critique of oratory in setting up the premise of his plea to revert the decision:

> I have no fault to find with those who propose to reopen the debate about Mytilene, and I do not support those who object to reconsidering matters of real importance several times over. On the contrary, the two things I consider most prejudicial to good counsel are haste and high emotion: the latter usually goes with folly, the former with crude and shallow judgement. As for words, anyone who argues seriously that they should not guide our actions is either stupid or has some personal interest at stake: stupid, if he thinks there is any other way to explore the future in all its uncertainty; self-interested, if he wants to argue some discreditable case but concludes that though he cannot speak well enough to carry a bad cause he can slander well enough to intimidate both the opposing speakers and the audience. (Thucydides, 42.0–42.2)

Here, Diodotus makes two key points: first, that important matters should be discussed thoroughly; and, second, that rhetoric is the only means of discussing 'the future in all its uncertainty'. These claims set the scene for Diodotus' intervention, which is posited in implicit opposition to foolish haste and coarse passion – and in support of careful consideration of all aspects of the issue, using persuasive speech as the means of scrutiny.

That day in Athens, as happens throughout Thucydides' writings (Conley, 1990, pp 2–3), the winning argument not only made a particular political case but also mounted a general defence of the art of rhetoric. Thus, Diodotus succeeded in persuading the assembly that only those individuals who had been directly involved in the revolt should be sentenced to death, whereas everyone else should go free. As a ship had already been sent out to announce the first decision, a second one was now dispatched, arriving just as preparations to carry out the first order were being made, but in time to halt the action. As Thucydides laconically concludes: 'So close did Mytilene come to disaster' (49.3).

Deliberative rhetoric: deciding for an uncertain future

Another conclusion that may be drawn from the story is that in Athenian antiquity the power of rhetoric was, indeed, great. Charlotte Jørgensen (2002) argues that Thucydides' account of the Mytilene debate offers a paradigm for deliberative rhetoric as envisioned in ancient Greece, which is not to be confused with the Habermasian ideal of deliberation/deliberative democracy (as presented in Chapter 3 and discussed further next). Where Habermas equates deliberation with a particular form of rational debate in which the public interest is the only legitimate stance, deliberative rhetoric encompasses all political communication, enumerating the different ways in which a rhetor may seek to persuade an audience to pursue a certain cause of action.

The case of Mytilene is paradigmatic in three key regards: first, it begins from what Jørgensen terms a foundational rhetorical situation in which a decision must be made between opposed views (change or maintain initial sentence); second, it proceeds through arguments for and against the proposal (the two speeches); and, finally, it ends with a vote (narrowly won by those favouring the changed sentence) where consensus is not reached, but the majority position is accepted even as disagreement persists.

Here, as Jørgensen also makes clear, it is important to reiterate that the speeches Thucydides presents are not historically accurate, but products of his own invention; they reflexively present the Athenian ideal of deliberative rhetoric as Thucydides envisioned it. Or, rather, Cleon's speech is the antithesis to this ideal:

> It works to subvert the possibility of an effective democratic rhetoric as Thucydides conceives that possibility. Cleon's rhetoric corrodes civic deliberation not only because it promotes narrowly partisan ends but, more importantly, because it casts suspicion on any appeal to common interests. (Leff quoted in Jørgensen, 2002, np)

Diodotus may be guilty of some of the same charges since he also appeals to expedience rather than to what is universally right. However, as Jørgensen shows, Diodotus' speech empowers the audience to independently make up their own mind. Diodotus argues in favour of a position that may be contested, and he does so in a manner that some will find contestable. But that is exactly the point: the encounter with contingent arguments enables the audience to exercise their own reason on the matter at hand, just as they may reflect upon the quality of the arguments with which they have been presented.

As such, the speech establishes and adheres to an ideal of deliberation in which there is no higher truth than that arrived at in and through deliberation itself. Diodotus may aim to persuade the audience to follow his advice, but in doing so he accepts the fundamental premise that arguments exist for and against different positions – or, more precisely, that in politics no position is singular or infallible; on the contrary, alternatives always exist, and every available option is somehow flawed. As we are dealing with matters of the uncertain future, there is no truth, only decisions (Kock, 2009).

Further, the speech explicitly favours ongoing deliberation of a matter; even after a decision is reached, it may still be divisive, subject to renewed articulation of disagreement rather than sunk into the silence of consensus. Or, rather, this is Diodotus' argument for reopening the case – once the second decision was reached, he would probably suggest that it was final had he been challenged to reopen it. Still, continued contention is the deliberative ideal that Thucydides illustrates with the account of the reversal of the decision, which remained divisive and was only passed by a small majority but was still reached in accordance with the codified procedure for collective decision making as inscribed in the Constitution of Athens at the time.

Thus, Thucydides illustrates that democracy rests on disagreement rather than consensus. Further, he shows that rhetoric is the means of expressing disagreement and, hence, is synonymous with democratic deliberation (understood as discussion of matters regarding which a decision must be made). Persuasion rather than truth is the outcome of this process.

This view of rhetoric and of deliberation has been contested throughout history and remains contentious to this day. Thus, it bears very little resemblance to policy making in modern societies, which tends to be a much more bureaucratic and technical affair (Miller, 2005). In fact, we might argue that the elected members of present-day democratic institutions, who are the modern counterparts to the members of the ancient Athenian assembly, tend to make up their minds first and then go looking for arguments with which they can defend their position from public scrutiny. For instance, this seemed to be the case at the level of the UN Security Council when in 2003 Colin Powell, the US Secretary of State at the time, laid out the case for an invasion of Iraq, claiming that Iraq possessed weapons of mass destruction and supporting the claim with false evidence (Borger, 2021). Or, going further, we might posit that today's politicians do not even bother to discuss their positions, but simply present contingent policies as accomplished facts. As Dean concludes in her analysis of the public communication of the decision to invade Iraq:

> Despite the terabytes of commentary and information, there wasn't exactly a debate over the war. On the contrary, in the days and weeks prior to the US invasion of Iraq, the anti-war messages morphed into so much circulating content, just like all the other cultural effluvia wafting through cyberia. (Dean, 2005, p 52)

Diodotus, let us be clear, argued strategically, presenting the arguments that would most strongly support his position, but he had a direct opponent and his audience made up their minds independently. Rather than forming the basis of contemporary democracies, this 'foundational' rhetorical situation seems to be something of a rarity today.

From a rhetorical point of view, this is lamentable, for, as Thomas Conley points out, rhetoric 'is about nothing if it is not about controversy' (1990, p ix). We might extend this argument to suggest that democracy is also nothing if not controversial – and, hence, rhetorical. Conley ties this point to the historical development of the art of rhetoric, suggesting that acts of persuasion become 'particularly important to people during times of strife and crisis, political and intellectual'. Thus, it is especially problematic that in today's climate of democratic polycrisis (see Chapter 1), the rhetorical notion of controversy is not taking centre stage.

In this chapter, I will move back and forth between classical rhetorical theory and modern theories of democracy to gradually build the argument that controversial encounters should be a foundational democratic principle – and practice. For now, it seems high time to return to the present to establish the continued relevance of *controversia* and begin carving out a conceptual space for it.

Between deliberative democracy and agonistic pluralism: situating *controversia*

Thucydides' account of the Mytilene debate illustrates a particular understanding of the role of rhetoric as the vehicle of democratic debate, which favours what Kenneth Burke (1945) calls 'the human barnyard', getting into the muck of discussing what should be done in particular situations, over and above philosophical dialogue about what is universally right, good and true. This view is very different from that of modern theories of the public sphere, as established in Chapter 3, where individual citizens model themselves to the idea(l) of the public as presented in texts that are addressed to this public, underscoring the reflexive and self-organized character of modern publics (Warner, 2002).

Conceptually, such individuated publicness, an imagined subject position to which the reader responds when reading the text that is addressed to this subject, is matched by the ideal of consensus. This is what Jürgen Habermas (1998b, p 306) describes as 'the unforced force of the better argument': when arguments are 'properly' addressed to the ideal of rational citizens, those citizens will understand them 'properly' as well, forming common opinions in and through their reasonable reception of arguments that are tailored to the general public rather than to any individual situation or particular audience.

More specifically, the conditions for such opinion formation are summarized in the 'ideal speech situation' in which everyone is free to forward any claim, which can then be tested as to its truth, validity and sincerity – and against a background measure of understandability (Rasmussen, 2019). The previous chapter focused on this modern concept of the public sphere, considering whether and how its ideal of consensus might apply to current digitalized practices of dialogue and debate. Here, I turn to classical understandings of the means and ends of the public exchange of opinions as these are inscribed in the theory and practice of rhetoric generally and the notion of *controversia* more specifically. I will suggest that the rhetorical approach might be a better candidate for explaining *and* intervening in current realities than that of deliberation.

The move from debate that aims at agreement to interactions that spur continued controversy entails a shift from a harmonious ideal of society to one which emphasizes that society should not only be able to accommodate differences, but may, in fact, thrive on conflict. This makes it all the more fitting that the opening example of this chapter discusses what to do in a situation of war. War (and peace) is a central substantial *topos* of classical rhetoric, that is, a topic of public concern, discussed on a regular basis and in recognizable ways. Aristotle (1359b), for instance, advises speakers on all the different things they must know about their own and other cities so as to be able to deliberate on matters of war and peace, listing historical knowledge about wars waged, the present probability of war and the likelihood that, were war to break out, one party would be stronger than the other as some of the most important aspects to consider.

Further, war has remained a controversial topic of political debate throughout history. And today, it is an urgently pressing matter, at the top of the political and public agenda of Europe once again, as the wars in Ukraine and in Gaza have brought armed conflict 'home' to Europeans. With such a statement, of course, one ignores the many ongoing conflicts around the world that are just as atrocious yet are not 'top of mind' for European (or 'Western') audiences. My point here is not to condone the privileging of some armed conflicts over others, but to note that this privileging takes place; some wars gain more attention – are felt more intensely – than others, depending on one's position in relation to them. As this chapter unfolds, I will attend recurrently to debates on (the prospects of) war. Let me reveal my persuasive intentions here: these examples are not just offered as illustrations of conceptual points, but in an effort to destabilize the privileging of some armed conflicts (and some positions within them) over others and to articulate the basic unjustifiability of war.

Indeed, war may be the continuation of politics by other means, as stipulated in Claus von Clausewitz' famous aphorism. And it may be (mostly) true that two democratic states have never waged war on each other as proponents of 'democratic peace' theory claim (Mello, 2017). When considered alongside each other, these two observations suggest that politics is also the continuation of war, indicating that conflicting opinions may clash without resolution. This position is implied in the many metaphors and analogies that liken rhetoric to war, talking about 'winning' debates, 'firing off' arguments, exposing the opponent's 'weak flank' and so on (Ritchie, 2009). However, that does not mean war is ever justified, and in this chapter I will consider whether and how we can conceptualize and practise rhetorical disagreement as an enabler of peaceful coexistence. Thus, I will argue that while controversial encounters do not lead to consensus, that does not make them synonymous with wars on words. Controversy, as I define it, is not about thwarting opponents to end discussions, but is concerned with the inclusion of different positions in ongoing debates.

The principle that democratic legitimacy should be based in sustained disagreement rather than achieved harmony (whether through consensus or by silencing the opposition) implies that the aim of public debate should be to maintain differences rather than resolve them. This is the core of Chantal Mouffe's (2013) theory of agonism. Democratic societies, Mouffe argues, are held together precisely because irresolvable differences continue to go up against each other rather than retreat into each their corner. In the absence of 'perpetual peace', we should perpetuate discussions of the principled justifiability of war as well as the justification of specific acts of war – and of interventions in (the debate about) war. Never should we reconcile ourselves to the inevitability of war – or, indeed, resort to violence, which (at least for those speaking against the justifiability of war, whether in principle or as specific acts) would defeat the purpose of debating war in the first place. As Mouffe says, invoking Marcel Mause, the 'aim is to "oppose without killing each other"' (2022, p 28).

While speaking up against war may, in principle, be an easy position to take, it is complicated by a myriad of practices. First, opposition to war is often a marginalized stance that involves resistance to 'the military-industrial-complex' and 'state-sanctioned violence' as they operate within one's own national context. Here, protests against interstate military conflict and international interventions are inextricably linked to domestic conflicts, and the entanglements and alignments of those who suffer 'at home' and 'abroad' make controversies of war particularly ingrained, woven into the fabric of society.

Second, a principled opposition to war may come into tension with one's support for those who suffer the consequences of war and/or one's desire to extend solidarity to everyone, including those who might be perceived as the aggressor in a conflict. We see this tension play out in controversies surrounding the war in Ukraine where most public commentators denounce the Russian invaders, but where disagreement persists as to the best means of supporting the Ukrainian people (Oltermann, 2022), just as one may be as concerned with the suffering of Russian civilians as with Ukrainians (Marimon, 2022) or take issue with the capitalist exploitation of Ukrainians that may follow from the war (Žižek, 2023). Even more contentious issues arise regarding the war between Israel and Hamas; here, it seems almost impossible to voice support for the Palestinian people without immediately being heard as denouncing the state of Israel. As Butler writes in a reflection on why it is so difficult to discuss the things 'that most urgently need to be discussed':

> One bumps up against the limits of a framework that makes it nearly impossible to say what one has to say. I want to speak about the violence,

> the present violence, the history of violence and its many forms. But if one wishes to document violence, which means understanding the massive bombardment and killings in Israel by Hamas as part of that history, one can be accused of 'relativising' or 'contextualisation'. We are to condemn or approve, and that makes sense, but is that all that is ethically required of us? (Butler, 2023)

Third, principled concerns with the correct response to war may be displaced by more immediate feelings and needs, as is arguably what happened when the war in Ukraine occasioned Finland's and Sweden's applications for membership of the North Atlantic Treaty Organization (NATO) – while Ukraine's application is a more direct response to the invasion and, hence, may be interpreted as the Ukrainian government's attempt to respond 'correctly'.

The various complications to the principled stance against war do not diminish the need to continue articulating it. We will never reach consensus (nor, perhaps, will we end war), but that is exactly why we should continue voicing our different opinions. While one might argue that this is also the deliberative stance, Mouffe explicitly pits the ideal of agonistic pluralism against that of deliberative democracy; as one favours conflict and the other consensus, they are also each other's opposites (Mouffe, 1999, 2017).

Mouffe's position has both been criticized for diverging from the ideal of consensus (Erman, 2009) and for continuing to rely on it (Knops, 2007), but I accept the difference between and validity of both positions. However, I suggest that neither consensus nor conflict is sustainable in its pure form – lasting consensus is an unrealistic ideal (and would, in fact, not be ideal at all, as it could only be attained by suppressing differing views), and perpetual conflict is practically untenable, as we need workable resolutions to many urgent matters for society to function. Thus, we need to mediate between deliberation with its aim of consensus, on the one hand, and agonism with its aim of disagreement, on the other hand – a role for which I posit the classical rhetorical concept of *controversia* as a good candidate.

Aiming at persuasion, *controversia* acknowledges consensus in the form of temporary agreement and accepts conflict as an underlying condition of possibility of all political debate – if we did not disagree, there would be no need for discussion. Suspended between these extremes, *controversia* offers a third conceptual space from which to understand how decision making becomes possible in practice; how we may come to agree on specific points even as we continue to disagree more fundamentally. Conversely, it helps explain how debates may continue even though a decision has been made. Full consensus is not a necessary precondition of legitimate political action and nor is full disagreement a prerequisite of legitimating public debate.

Seeking to substantiate *controversia*'s candidature as conceptual mediator, I will discuss the classical rhetorical concept in relation to modern theories and practices of controversy. Further, I will position articulations of controversy in relation to other modalities of disagreement, suggesting how the concept of *controversia* may invoke a distinct stance, but should nevertheless take inspiration from more reconciliatory as well as more escalatory approaches. *Controversia*, then, is the theoretical concept and controversy is the empirical practice that I posit as the ideal for democratic debate – a flawed and provisional ideal, but, I argue, better than anything else we've got. To explore both the theory of *controversia* and practices of controversy, I will conduct a 'vertical' (that is, historical) reading of the concept and a 'horizontal' (that is, contemporary) reading of its current modes of articulation, thereby seeking to reclaim a space for persuasion between consensus and conflict. But first a caveat is in order.

A note on (de)contextualization

When excavating a historical concept for current use, it is important to recognize the limitations of this manoeuvre – especially when it is conducted for normative purposes, as is the case here. The classical concept of *controversia*, which inspires my suggestion for 'making disagreement good again', comes with a hefty and heavily problematic baggage. Thus, it is paramount that we dislodge the concept from its historical context, separating the idea from the system that enabled its development. More specifically, the concept of *controversia* must be freed from any and all economic structures that are based on enslaved labour and separated from political systems in which women and other historically disenfranchised groups have no say. And this reconceptualization must be supported by – and give support to – practices of controversy that actively encourage equal and inclusive societies. The concept of *controversia* that I develop is inspired by classical rhetoric and its iterations, but the theoretical ideal it posits has very little to do with historical realities.

To open up the conceptual frame of *controversia* to alternative articulations of controversy, the position I develop here emerges from a concerted postcolonial and feminist critique of the concept's classical roots as well as its current branches, a critique that radicalizes the sensitivity to temporality and positionality, which is inherent to the notion of *controversia*. 'Arguing from *all* sides' is the means *and* end of this revamped notion. To support such conceptual work, I will draw on empirical disagreements that involve asymmetries of power as well as the attempt to overcome historical and contemporary injustice. While I have so far discussed the politics of war, I now turn to conflict, more broadly, focusing on efforts that aim at opening up the space of public disagreement to more positions and voices. Thus,

I seek alternatives to armed conflict and repressive consensus – whereas both seek to overcome disagreement, whether by the sword or the word, the aim of *controversia* is to build and maintain relations of difference.

Arguing from *all* sides

The question of what *controversia* means, conceptually speaking, is part of a broader debate within rhetoric, both classical and modern, on how to define and practice persuasion. Conley (1990, pp 23–24) distinguishes between four basic models of rhetoric that he traces back to the Greek roots of the art of speaking, associating each model with one of the art's 'founding fathers'. Gorgias is posited as an exponent of *motivistic* rhetoric that is mainly concerned with swaying the audience at all costs and sees emotional appeals as holding more sway than rational arguments. This is the model of rhetoric that has, subsequently, given the art a bad name, raising accusations of a lack of substance and a manipulative take on form.

Plato, for one, raised this critique, most notably in the dialogue that bears Gorgias' name and offers an out-and-out rejection of rhetoric. In other dialogues, most famously *Phaedrus* but also in the dialogue named after Protagoras, Plato has Socrates define an alternative to Gorgias' and other sophists' approach to rhetoric, establishing the model of *dialectical* rhetoric that is only practised in the service of 'Truth' with a capital T – a rhetoric that aims at clarity and uses reason to lead the audience to the right conclusion. This view of rhetoric can be criticized for its assumption that the speaker knows the truth and leads the audience to it, which privileges the speaker and ignores the many cases in which 'small t' truths are the only available and viable positions, just as it misses the point that rhetoric is about deciding on what to do rather than on what is true.

Aristotle, who can be said to combine the motivistic and dialectical views on rhetoric, is the key representative of what Conley terms the *problematic* model of rhetoric. Here, the focus is on the available means of persuasion in any given case, indicating that these will vary according to the problem at hand and the situation in which the problem is addressed. Aristotle carves out a niche for rhetoric that avoids sophistry by attending to the 'available means' rather than persuasion itself, while insisting that this attention is valid. Thus, he is not concerned with persuasion at all costs, but with finding out what persuasion might look like in the specifics. In this regard, the problematic stance is also analytical.

By acknowledging that the means of persuasion change from case to case, Aristotle nuances and broadens Plato's dialectical position. In Aristotle's view, there is no one way in which audiences can be let to realize the truth, but many different ways in which the rhetor can attempt to shape various audiences' opinions. The means of persuasion, then, may be generalizable in some respects, but they are always available in particular forms, depending on the purpose of the speaker (that is, the genre of the speech as well as its aim) and the constitution of the audience (for example, their preceding knowledge and opinion of the case). Significantly, this view leads Aristotle to delineate the subject area of rhetoric to the matters about which deliberation is necessary because they are uncertain and because something can be done about them. 'For no one', he asserts, 'debates things incapable of being different either in past or future or present' (1357a).

Rhetoric, in the Aristotelian model, is about 'proposals, not propositions', a view that is today prevalent in theories of practical argumentation that seek to retrieve the notion of deliberation from deliberative democracy and return it to its rhetorical roots (Kock, 2009). Here, deliberation is associated with politics in and through the genre of deliberative rhetoric, as defined by Aristotle – that is, as proposals about how to act collectively where the action is directed towards the future and the outcomes are uncertain. Recovery of Aristotle's understanding of deliberation specifically and of rhetoric generally has been pivotal to 20th-century resurgences of the art of persuasion, not least because of its analytical properties and its pragmatic view (Gaonkar, 1993). I will return to the issue of the modern uptake of classical rhetoric shortly, but first let me present the fourth and final model of rhetoric, which was, according to Conley (1990, p 20), the most pervasive in the classical context.

Conley associates this model, the controversial view of rhetoric, with two Greek rhetoricians, Protagoras and Isocrates, but he finds its main exponent in Cicero, the Roman politician, lawyer and rhetorician. Incidentally, Conley (1985) also champions the controversial view, offering a modern interpretation of the classical concept. In terms of the classical roots, the Latin term *controversia* finds its Greek antecedents in the notion of *dissoi logoi*, the idea, as expressed by Protagoras, that there are arguments for and against any proposition. This is also referred to as *antilogic* or the building of arguments on either side of a case. Protagoras, whose ideas are only available to us through fragments and secondhand retellings, is associated with the claims that humans are the measure of all things and that it is always possible 'to make the weaker argument stronger', which may be taken to indicate that the measure of things can be warped, leading to rampant relativism. However, it may also be interpreted as a more temperate assertion that 'only by examining the arguments for and against a given proposition can one come to some determination as to which side is to be believed and acted upon' (Conley, 1990, p 5).

Hence, Protagoras' position can be defended as 'merely' suggesting that when humans are to judge in a matter, it is prudent to consider all the aspects and dimensions of that matter (the arguments for and against) so as to become better able to decide what is the best position and the right thing to do. This situates *controversia* as a basic prerequisite of democratic societies, as the foundation for the exchange of differing views between citizens with equal rights. As Michael Mendelson notes:

> Antilogic [Protagoras's term for *controversia*] is on the side of radical egalitarianism: we are all endowed with something to say, with the 'truth' as it appears to us, and in the process of struggling towards a shared or consensual view, it becomes an 'inalienable right' for all positions to have the floor, if not to carry the day. (2002, p 53)

This 'egalitarian' sentiment was carried forward by Isocrates and into Roman rhetoric, forming the basis of an entire educational programme. Codified by Cicero in his vision of the 'perfect orator' as someone who is able to consider a matter from 'all sides' ('in utramque partem') and echoed in Quintilian's ideal of the 'good person, skilled in speaking' ('vir bonus dicendi peritus'), Isocrates pioneered the idea that 'the ability to speak well and think well go hand in hand' (Conley, 1990, p 18). Thus, engaging in controversy became established as the legitimate means to deciding on societal ends, and cultivation of persuasive skills was, for centuries, the main benefit of having received what we might today term a classical education. Beneficiaries of this educational programme became able to exercise their 'inalienable right' to speak up for their own version of 'truth' – or, rather, to express their opinion on all matters subject to collective decision making. This already hints at one way in which the model, as originally conceived, was less egalitarian in practice than in principle, only being available to those who had the right gender and class for such schooling. However, before turning to the critique and reconstruction of *controversia*, let us look at its congenial modern reception.

Reviving *controversia*

While the controversial model of rhetoric has lived on throughout the centuries, it has, as will be seen in the next section, lived in the shadows of dialectical and motivistic models, respectively. Further, it was not central to the 'revival' of rhetoric in the late 19th and early 20th centuries, which was, instead, spearheaded by so-called neo-Aristotelian approaches, focusing on the analytical process of assessing whether and how speakers make appropriate use of the available means of persuasion in particular circumstances (Enos, 2006). The Aristotelian inspiration, Dilip Parameshwar Gaonkar argues, has enabled modern critics to posit rhetoric as a 'hermeneutic metadiscourse':

> The fact that the ancients had a sharply different estimate of theory than we do frequently escapes our attention because our understanding of rhetoric – both its history and its current possibilities – is mediated almost exclusively by Aristotle rather than by Cicero and Quintilian. Aristotle's text, perhaps uniquely in the entire Greco-Roman rhetorical canon, permits us to entertain a distinction between theory and practice, with the former in some sense regulating and rationalizing the latter. (Gaonkar, 1993, p 260)

While this distinction enables modern critics to apply classical rhetorical concepts to the analysis of current rhetorical practices, Gaonkar asserts that it leaves them blind to the ways in which present practices differ from those of the past, just as it deflects attention from the interrelations of theory and practice. As such, he concludes, we should develop theories that are more attentive to changing rhetorical practices.

Such attention, I suggest, is facilitated by the controversial model of rhetoric. Even if the irony of answering the critique of one classical model with the offer of another should not be lost on anyone, Gaonkar does indicate that it might work in the case of rhetoric. Thus, he highlights how theory and practice have been considered conjointly in the rhetorical canon, associating this conjunction with Cicero and Quintilian. Here, the *concept* of *controversia*, as directly inspired by Isocrates and more implicitly drawing on Protagoras, is intimately linked to the *practice* of controversy, with the latter informing the former rather than the reverse. Thus, the controversial model offers us a means of theorizing rhetorical practice in and through the study of specific practices. It does not provide 'cookie cutters' with which any and all rhetorical efforts can be judged against classical standards (Willis-Chun, 2008), but sensitizes us to the interrelations of rhetorical situations and rhetorical responses, to the ways in which 'the means of persuasion' are both made available by and constitutive of the circumstances of rhetorical in(ter)vention.

As was indicated earlier, Conley is one prominent exponent of the (re-)appropriation of classical *controversia* for the study of modern controversy. Taking his cue from Richard McKeon, Conley argues that 'it is necessary to accept the plurality of legitimate and coherent lines of argument' (1985, p 470). In other words, we should approach persuasive processes with the assumption that controversy will persist even after a decision has been reached. *Controversia*, Conley explains:

> requires that both sides of any question be heard and that thereby, within the resulting disputation, invention and judgment of truth or verisimilitude be obtained. Ciceronian debate does not move platonically from communities and oppositions to an assimilation of lesser truths to the greater; it is resolved only insofar as anyone chooses

> to adopt one of the positions in conflict or to modify it or to formulate a new position using elements from the alternatives as components. (1985, p 471)

Thus, the end of controversial encounters is not consensus, but rather the informed choice of the parties involved, with everyone being sufficiently persuaded by one view or another to be able to cast their vote in favour of it, or seeing enough merit in each view to begin developing compromises (or, indeed, finding so little value in either as to start thinking of other alternatives).

This is the model by which the Athenians were able to decide what to do in the case of Mytilene and, hence, the ideal presented by Thucydides in the story with which I opened this chapter. And, arguably, it is still the ideal of Western democracy. For instance, drafts of the constitutional treaty of the EU, which was agreed by the political leaders of all Member States in 2004, but later failed to be nationally ratified, contained a key quote from Thucydides long after a reference to the Christian heritage of Europe had been removed from the developing document (Orgad, 2010). Using Thucydides' rendition of Pericles' funeral oration to suggest that 'our Constitution … is called a democracy because power is in the hands not of a minority but of the greatest number' was deemed less exclusionary than reference to religious allegiances. Still, the Greek epitaph was not carried into the final document, nor did the constitution for Europe ever become a reality, suggesting that the historical roots and current character of (European) democracy remain disputed.

Conceptually speaking, we might praise the notion of *controversia*, as developed here, for assuming that dispute is a constitutive principle of democratic societies rather than an obstacle to be overcome (for example, through recourse to 'the greatest number'). However, we may also raise the initial objection against the classical concept that it assumes a dispute only has two sides. If the concept is to be open to changing practices and, hence, amenable to conceptual change, it must be able to look beyond this dialectic. Further, its reconceptualization should involve not only the observation of societal change, but also a more foundational critique of the societal context in which it emerged – and of the social norms it continues to uphold. Before getting to the reconceptualization of *controversia*, however, let's have a brief look at the history of rhetoric. While covering the two millennia that separate the classical concept of *controversia* from its contemporary proponents in a few pages will clearly not amount to an adequate conceptual history, it will establish the plight of rhetoric: its uneasy relationship with truth.

The plight of rhetoric

A simplified account of the history of rhetoric would suggest that whereas the controversial model of the art was influential in antiquity, it has subsequently been submerged in the controversy(!) between dialectical and motivistic models of rhetoric. Here, proponents of the former model have held out the latter as exemplifying the kind of superficial, shallow or 'merely' ornamental practice that should be abandoned in the pursuit of higher truths – and of a rhetoric able to serve such truths.

From Plato onwards, much has been made of the fact that, like medicine, which can also be poison, rhetoric can be used for both good and evil. In the dialogue *Gorgias* Plato extends the analogy to also include the relationship between medicine and cookery:

> Cookery simulates the disguise of medicine, and pretends to know what food is the best for the body; and if the physician and the cook had to enter into a competition in which children were the judges, or men who had no more sense than children, as to which of them best understands the goodness or badness of food, the physician would be starved to death. A flattery I deem this to be and of an ignoble sort. (Quoted in Craig and Muller, 2007, p 118)

Here, rhetoric is, at best, flattery and, at worst, poison. And whereas medical doctors can resolve the tensions of their discipline by taking the Hippocratic Oath, the challenge of avoiding useless superficiality and harmful manipulation remains foundational for rhetoric. This leads many thinkers, including Plato himself, to advocate a rhetoric that begins from and only speaks for a pre-existing truth (as identified by the speaker in advance of speaking).

Christian rhetoric and, subsequently, the rhetoric of Enlightenment offer two versions of the dialectical model of rhetoric, which put the art in the service of the truths of 'God' and 'Science', respectively. Notably, Augustine believed that if heathens were making effective use of rhetoric in the service of their falsehoods, Christian preachers should also be allowed to 'embellish' the truth. Unlike Aristotle, who asserts firmly that 'the true and the just are by nature stronger than their opposites' (1355a), Augustine seems doubtful of the power of truth to make its own case compellingly:

> Now, the art of rhetoric being available for the enforcing either of truth or falsehood, who will dare to say that truth in the person of its defenders is to take its stand unarmed against falsehood? For example, that those who are trying to persuade men of what is false are to know how to introduce their subject, so as to put the hearer into a friendly, or attentive, or teachable frame of mind, while the defenders of the truth shall be ignorant of that art? (Augustine, Book IV, Chapter 2.3)

Augustine was trained in and taught rhetoric before converting to Christianity, and, arguably, rhetorical thinking saturated his theology beyond the defence mounted here – that is, beyond being a legitimate means to the end of preaching God's truth, rhetoric may have shaped how Augustine conceived of that truth (Gronewoller, 2021).

Contrary to Augustine's more conciliatory view, exponents of scientific truth tend to reject the communicative and intellectual legitimacy of rhetoric. Thus, Enlightenment philosophers associate rhetoric exclusively with vain flattery and suggest that their exposure of truth is devoid of such embellishment, relying instead on logical thinking and the 'pure' representation of thought (Conley, 1990, p 191). Here, rhetoric becomes synonymous with aesthetics and expelled from the worlds of science and politics, stripped of its formative role in the public sphere and reduced to the realm of (guilty) private pleasures.

This historical account of rhetoric as living in the shadows – and at the mercy – of the Christian Church and/or the Church of Science ignores the rhetorical richness of the renaissance and hardly does justice to medieval rhetoric either (Murphy, 1974). However, it does accurately reflect the plight of rhetoric, variously positioned as an auxiliary to other disciplines and/or associated with dubious practices as it has been throughout history (Vickers, 1989). Therefore, to this day, rhetorical scholars feel the need to answer questions about the value of persuasion and to address concerns about the ways in which people are persuaded, rehashing a fundamental tension between the ethical and the effective use of rhetoric that, again, harks back to Plato and still compels rhetoricians to defend themselves and their field (Garsten, 2006).

The defence of persuasion leans into broader philosophical discussions of whether language is a 'mere' medium and (potential) obstacle for a truth that exists outside of speech or a formative force of social reality and individual experience (Lash, 2015) – or, as we might ask today, is language descriptive or performative? Here, rhetoric took the side of the performative millennia before the 'linguistic turn' in philosophy. People do not carry around bags, whether literal or metaphorical, from which to draw representations of their experiences in the world; rather, we form those experiences when speaking of them.

This pragmatic view has been the 'ground truth' of rhetoric since its inception, and it now seems a quite uncontroversial starting point across the disciplines interested in processes of meaning formation. Whereas rhetoric's insistence that 'you can do things with words', millennia before the introduction of speech act theory, has marred its history, we can now tell the gratifying tale of how rhetoric was right all along (Isager and Just, 2005). Thus, interest in the art of persuasion has risen in step with the increasing recognition of the performative powers of speech.

Rhetoric, check your privilege

Still, something is rotten with the state of rhetoric. For one thing, the seeming triumph of performative over representative theories of speech has not quite led to the celebration of persuasion that one might expect. I will develop this point in the next two chapters, detailing how current sociotechnical developments have incurred what I term the revenge of transmission theories of communication and led to the closing of the rhetorical mind. Further, and as I will discuss here, it may be that rhetoric can successfully position itself as an underdog of intellectual history, but that is still a very privileged position – and one that is built on the systematic erasure and exclusion of other subjects, whether understood as topics of concern or human beings.

If we are to recover the rhetorical concept of *controversia* for present use, then, we must begin by recognizing who the concept is built by and for; we must check its privileges. As Jacqueline Jones Royster points out:

> Western rhetorics, at least the legacies of them that we have inherited through scholarship, are demonstrably dominated by elite male viewpoints and experiences. Twenty-five hundred years of rhetorical scholarship … are, in fact, testament of Western dominance in interpretive authority and of the situating of that authority in male-dominated and elite ways. (2003, pp 149–150)

Raka Shome makes the point even more emphatically: 'Rhetoric as a discipline is largely based on humanist theories and speeches of white men in power and has not been adequately self-reflexive about its scholarship in relation to issues of race and neocolonialism' (1996, p 49). Given that the privileging of White, male speakers runs through the theory and practice of rhetoric, we might position the art of persuasion as one of 'the master's tools' that Audre Lorde (1981) suggests 'will never dismantle the master's house'. This raises the question of whether and how it is possible to use rhetoric for emancipatory purposes.

One way of substantiating an affirmative answer to this question is to turn to the ways in which rhetoric has, historically, been thought and practised by those who are written out of dominant accounts of the rhetorical tradition(s). Thus, it is possible to write the history of rhetoric differently, extending the disciplinary scope beyond the different models of rhetoric that have struggled to define the field to also attend to the versions of rhetoric that have been silenced in mainstream accounts (Miller, 2008). Dominant accounts notwithstanding, rhetoric was never an exclusively Western or an exclusively male phenomenon. On the contrary, there are rich traditions of rhetoric across the globe with quite different understandings of what persuasion might be and how it could be practised (Donawerth, 1994). And, if one cares to look, there are women rhetoricians throughout history, influencing both the theory of persuasion and practising persuasion to influence their world (Biesecker, 1992; Lunsford, 1995).

Building a feminist and postcolonial theory of rhetoric, then, may begin by recognizing the dominance of the field's male and Western heritage, but will immediately diversify our understanding of persuasion by emphasizing ways in which it has been thought of and practised outside of the White, male gaze. In other words, this perspective decentres the version of theory that otherwise tends to think of itself as 'the view from nowhere' (Haraway, 1988), showing how every theoretical perspective is just that – a perspective. The question is therefore not how to avoid perspectivism, but which perspective we take, just as we must consider what our perspective excludes in and through its inclusions.

For the concept of *controversia*, its classical iteration shares the exclusionary mechanism that Jacques Rancière identifies in the Greco-Roman notion of politics: 'The demos attributes to itself as its proper lot the equality that belongs to all citizens' (2004, p 8). For Rancìere, interrupting the illusion of equality constitutes politics proper; it is 'the initial twist that institutes politics as the deployment of a wrong or of a fundamental dispute' (2004, p 13). Here the twist is to expose the body of the demos as a specific body, thus freeing the excluded from the burden of having to cover up this fact. Politics, with this twist, springs from the productive tension of inclusion through exclusion, its purpose being to usurp the means of exclusion for the purposes of inclusion. In the classical context, Rancière suggests that the fundamental tension revolved around enslavement: 'The slave is the one who has the capacity to understand a logos without having the capacity of the logos' (1999, p 17). This exclusion sustains the citizens' illusion of inclusion – and the rebellion against it constitutes politics in Rancière's view.

A similarly fundamental exclusion was also at play around gender, with women (in ancient Greece, classical Rome and beyond) being able to hear the discourse of men, but not having the right to speak up in public. Just as Rancière recounts classical instances in which enslaved people were able to use speech to instigate and support rebellion, so there are classical examples of women speaking up to establish their right to speak. While few accounts survive, their very existence, whether as historical documents or in literary form, indicate that women's exclusion from the demos was as contentious a matter as the exclusion of enslaved people (Glenn, 1994; Jarratt and Ong, 1995; Saxonhouse, 2005). Thus, recognizing these foundational exclusions, both conceptually and in practice, offers a starting point for disputing the injustices they incur – and for changing the situation.

Butler suggests something similar in their account of how the inclusion/exclusion of grievable subjects is related to the circulation of discursive frames, emphasizing how the (re)production of inclusion by means of exclusion contains its own undoing:

> As frames break from themselves in order to install themselves, other possibilities for apprehension emerge. When those frames that govern the relative and differential recognizability of lives come apart – as part of the very mechanism of their circulation – it becomes possible to apprehend something about what or who is living but has not been generally 'recognized' as a life. What is this specter that gnaws at the norms of recognition, an intensified figure vacillating as its inside and its outside? (Butler, 2016a, p 12)

This, I believe, is the central question to ask of controversies: what is it that they exclude in order to establish themselves as controversies? What are the arguments that lie outside of what is recognized as 'the two sides' of the disputed matter? And how could these arguments gain recognition?

As Butler argues, the very reproduction and circulation of dominant frames constitutes the basis for altering their meaning and for breaking free of their normative confines:

> The movement of the image or the text outside of confinement is a kind of 'breaking out', so that even though neither the image nor the poetry can free anyone from prison, or stop a bomb or, indeed, reverse the course of the war, they nevertheless do provide the conditions for breaking out of the quotidian acceptance of war and for a more generalized horror and outrage that will support and impel calls for justice and an end to violence. (Butler, 2016a, p 11)

Thus, it becomes possible to move beyond the confines of controversies as currently constituted. In Butler's case, this involves displacing the internal controversies of war, refusing to answer the question of whether you are 'with us' or 'against us' in a fight between good and evil, as 'the war on terror' was (in)famously framed (see also Ivie, 2007). Instead, Butler suggests, we should question the very framing of these controversies, their external reliance on 'the quotidian acceptance of war', shifting the perspective from the question of choosing sides in (armed) conflict to the issue of how to resolve conflicts by nonviolent means.

To suggest how such reframing may evolve, we can return to Rancière's point that enslaved subjects enact politics in and through their rebellion. Today, we might suggest, democratic illusions of equality are not sustained by the operation that negates the enslaved body even as it performs the work of society. Yet, people continue to work under conditions of slavery across the globe (Whiteman, 2023), and this work seems to be more ingrained and less visible today than it was in the past. It is easier to call out slavery that acknowledges the label than it is to identify involuntary bondage to exploitative working conditions, which pretend to offer workers a choice. Further, the history of slavery continues to be controversial, as some countries are reckoning with their past while others still reject the need for historical reflexivity.

As a case in point, in 2023 the Dutch king offered an official apology on the occasion of the 160th anniversary of his country's abolition of slavery, but in Danish reactions to this news, leading politicians refused to consider that similar atonement might behove the Kingdom of Denmark. 'Should we also apologize for what the Vikings did?', one right-wing populist asked rhetorically (Prakash, 2023). Well, yes, we should. But the fact that such reckoning is presented as absurd indicates the boundaries of current frames of inclusion and exclusion in Danish public debate, suggesting where it might be possible to move 'the quotidian acceptance' around issues of structural racialized inequality and discrimination in the context of 21st-century Denmark.

While the question of inclusion and exclusion is, in some respects, a very tangible matter of ontological existence, it is intimately linked to the epistemological question of its discursive framing (Butler, 2016a). As Shome writes, 'whereas in the past, imperialism was about controlling the 'native' by colonizing her or him territorially, now imperialism is more about subjugating the "native" by colonizing her or him discursively' (1996, p 42). It is this ongoing colonization that we must recognize as a constitutive tension of rhetorical theory and practice:

> We simply *need* to engage in postcolonial analyses of texts. We *need* to develop critical perspectives that now seek to examine and expose to what extent neocolonial forces, whether they be representations of 'others' or representations of self, underwrite cultural, political, and academic discursive practices, for … if texts are sites of power that are reproduced by their social conditions, then neocolonial and racial forces are, to some extent, always already written into our texts. (Shome, 1996, p 51, emphasis in original)

The issue of inclusion and exclusion, then, is theoretical as well as practical; just as empirical controversies are constituted by their norms of recognition, so is their conceptualization. In order to sensitize the concept of *controversia* to the exclusions on which it is built and by which it continues to operate, we must identify the mechanisms of this exclusion.

Further, we must ask about, and become able to hear, silenced voices, finding ways of listening to what those whose speech is deemed meaningless are, in fact, saying. Sometimes silenced voices become quite palpable, as when

Chinese protesters in 2022 took to demonstrating their opposition to the regime with white sheets of paper (Murphy, 2022). Also, incomprehensibility can become apparent, as when in 2023 one of the two Greenlandic members of the Danish Parliament held a speech and answered questions in Greenlandic as part of a parliamentary debate on the Danish, Greenlandic and Faeroese commonwealth (Høj, 2023). We may see these protests and hear this voice, making it impossible to ignore their exclusion.

However, seeking to imbue invisible positions and/or silenced voices with meaning incurs the risk of essentializing, of colonizing the marginalized once more by representing them in a certain manner or appropriating them for one's own purposes. The representation of 'non-Western' and/or 'women's' rhetoric, as Mohanty (1984) forcefully shows, all too often resorts to sweeping claims on behalf of, for instance, 'the Third World woman', which one would never entertain for Western and/or male speakers. When wishing to represent positions that are different from our own, we should consider very carefully whether and how to do so.

Citing Spivak, Shome argues that essentializing is unavoidable and can be useful:

> Spivak suggests that while it is true that to engage in a postcolonial criticism that challenges the misrepresentations of racial 'others' in hegemonic discourses one does to a certain extent end up essentializing, that essentializing, however, is only a necessary 'strategic' essentializing – a risk that the critic *must* take 'in a scrupulously visible political interest'. (1996, p 47, emphasis in original)

The political interest of the present project is, as was established in the introduction, to 'make disagreement good again'. To this end, I harness a version of the classical concept of *controversia* that is, first, pluralized, moving disagreement beyond the dichotomy of for and against to include 'all' sides. Second, this concept of *controversia* begins from the theoretical premise that the ideal of full inclusion is impossible. Third, and finally, this implies a conceptualization that must constantly seek to renew itself by becoming aware of and attending to that which it disregards. It is to this final task that I now turn, seeking to detail a contemporary concept of *controversia* that is sensitive to present practices of controversy.

Contemporary controversies

Despite its centrality to classical rhetoric – and the concerted efforts of its modern proponents – the (re)conceptualization of *controversia* has not been a main concern of contemporary rhetoricians. In fact, as Kendall Phillips argues, 'the concept of controversy is mysteriously unexplored in argument studies and rhetoric' (1999, p 488). In other words, rhetoricians conduct case studies of controversies, but tend to use these empirical cases as occasions for exploring other theoretical issues than those that might be raised by attending to their very status as controversies.

However, within science and technology studies (STS), controversies have become a central research topic, especially among those STS scholars who are inspired by Bruno Latour's approach to mapping controversies. Here, emphasis is put on scientific controversies, but the identification and exploration of these begins from an understanding of controversies as exchanges of diverging points of view in the broadest sense:

> *Controversies are situations where actors disagree* (or better, agree on their disagreement). The notion of disagreement is to be taken in the widest sense: controversies begin when actors discover that they cannot ignore each other and controversies end when actors manage to work out a solid compromise to live together. Anything between these two extremes can be called a controversy. (Venturini, 2010, p 261, emphasis in original)

More specifically, Tommaso Venturini (2010, pp 261–262) explains, controversies are defined by five key features: (1) they involve all sorts of actors who (2) are involved in social dynamics around issues that (3) are resistant to any reduction of their complexity and (4) are articulated as and in debate, which (5) incurs the clash of opinions (that is, conflicts). Clearly, this definition is not delimited to scientific controversies, but rather could be said to have travelled from the field of politics into that of science, underscoring the different ways in which scientific knowledge can be controversial – as input to policy processes and, more fundamentally, when scientists themselves disagree about their knowledge claims.

The latter of these two types of scientific controversy has travelled back to rhetoric, as it were, through increased attention to the rhetoric of science, understood as the ways in which scientists seek to persuade their scientific communities as to the truth of their claims (Gross, 2008). The former has inspired a rhetoric of controversy that is particularly attentive to the ways in which personal, technical and public spheres of argumentation are delineated from and (re-)entangled with each other (Goodnight, 1982, 2012). What is at stake here is the interrelation between knowledge claims and political agendas, alerting us to 'the strategic instability of the distinction between epistemic and policy issues, between expert and public forums, between scientific and science-based controversy' (Miller, 2005, p 36).

To exemplify this instability, think of the different ways in which scientific knowledge (or the lack thereof) informed policy decisions throughout the COVID-19 pandemic (Hodges et al, 2022). Also, think of how the political attention to the pandemic influenced science, leading to, for instance, the fast development of vaccines (Forman et al, 2021). And think about how this scientific speed and political expediency has fed anti-vaccine conspiracy theories as well as more moderate vaccine scepticism, adding further strains to already pressured health systems and political institutions (Jennings et al, 2021). By attending to the interrelations of scientific-technical and public controversies, we may (re)discover the broader relevance of controversy for understanding processes of meaning formation.

For some rhetorical scholars, this leads to (re)instating controversy as constitutive of 'a contemporary public that is brought into being by oppositional argument[,] contests the privatization of common interests, challenges social conventions for justification, and works to expand a sense of shared interests to all affected by common action' (Olson and Goodnight, 1994, p 272). As such, these scholars argue, controversies may be seen to do 'the traditional work of the public sphere', but it should be noted that this work is limited to single issues and does not extend equally to all citizens.

Thus, controversies, even when they are temporarily resolved, do not lead to consensus – a fact that can be decried as the reason why controversies are democratically insufficient, if not directly suspect, or it can celebrated as their main contribution to democratic societies. Phillips takes the latter stance, suggesting that the value of controversies does not lie in their semblance to theories of the public sphere, but in their disconnection from such theories (and the practices they idealize):

> Disconnecting the notion of controversy from the theoretical tradition of the public sphere suggests embracing the disorienting fragmentation of contemporary society and relinquishing the normative space of the public sphere and grand conclusions of History in favor of the endless struggles enacted at the intersections of specific sites and momentary opportunities. (Phillips, 1999, p 492)

This implies localizing controversies, situating the articulation of specific issues as the main legitimizing force of democratic society (to which scholars of public debate, whether coming from rhetoric or STS, should devote their attention; see Marres, 2007): 'controversies must be understood as specific, localized intersections of discursive sites and momentary opportunities' that 'provide momentary opportunities to resist, change, and reform the local practices of those involved' (Phillips, 1999, pp 493, 495). While the potential of controversy to spark agreement – and the desirability of this potentiality – remains disputed (Goodnight, 1999; Ono and Sloop, 1999), the value of exploring specific articulations of controversy further – and of further conceptualizing such articulations – has been firmly established.

In his work, Phillips (1999, 2006) pays particular attention to the articulation of contradictions within established discourse formations, suggesting that such articulation may incur the reproduction of alternative discourses as well as the repositioning of the subjects that articulate them. Thus, he argues, we should focus on 'rhetorical maneuvers', on the specific ways in which subjects can make their voices heard, their positions known and, perhaps, their opinions matter in relation to issues with which they are concerned. This returns us to the question of who has the ability to partake in controversy – or, rather, to the issue of what such participation looks like and what subject position(s) it enacts. It alerts us to the entanglements of agency and subjectivity (Phillips, 2006, p 311).

One take on this issue likens participants in controversies to the figure of the 'parrhesiastes', as Foucault (2001) famously rearticulates the classical concept to construct a subject position of 'speaking truth to power'. However, this mode of 'fearless speech' persuades only with – and can only be persuaded by – the power of truth. Thus, the parrhesiastes tows the confrontational line and makes the 'hard case' for (and from) their own certainty. This can be an extremely brave stance, taken at the risk of the reputation – if not, indeed, the life – of the speaker. But it can also be a stance that serves to reinforce faultlines and harden conflicting positions, offering controversy but no potential for encounter. For instance, when Danish author Jonas Eika was awarded the Nordic Council Literature Prize in 2019, he gave a speech that criticized Danish immigration policy in no uncertain terms:

> I'm speaking to the Danish prime minister (who is also sitting in this hall). Mette Frederiksen, who is heading a social democratic party, which has gained power by adopting the former government's racist language and policies. Mette Frederiksen, who calls herself the children's prime minister, but conducts an immigration policy that splits families, makes them poor, and subjects children and adults to slow, degrading violence in the so-called departure centres of the country. (Eika, 2019, my translation)

While the speech was celebrated by Eika's political allies for its direct demand that the 'departure centres' be closed, it was all but ignored by incumbent politicians. In fact, the Prime Minister, to whom the parrhesiastic message was addressed, explicitly refused to respond to it (Ritzau, 2019). Regrettably, for the immigrants and refugees, to whom Eika sought to lend his voice, nothing has changed.

To enable change, the subject who engages in controversy must be open to the clash between ideas – and to the idea that the conflict could change them as much as it might change the opposition. As I seek to conceptualize controversial encounters, they are, indeed, encounters – meetings between different subjects in which each is positioned in relation to the other. Still, the subject who engages in controversies may be a provocateur who uses the confrontational style of polemics to gain the attention of the public, risking their alienation in the process (Rand, 2008).

Such provocations take many forms, one example being the wave of activism against colonial statues, which began in 2020 and has involved the toppling, destroying and spraypainting of contentious memorials around the world. This has sparked a reckoning with colonial history, but has also raised questions as to the appropriateness of the methods used (*New York Times*, 2020). Whether or not one agrees that the aim justifies the means, this seems to be the rationale of climate activists who, from 2022 onwards, have copied the tactic of painting over art works, focusing on famous works generally rather than targeting problematic pieces – the aim in this case being to shake members of the public out of their duplicity through spectacular activism that gains media attention (Sparrow, 2022), and the risk being that the aim is sidestepped in discussions of whether or not the means were justified, drawing attention to the activism, but away from the cause.

Polemic activism may be an efficient means of gaining public attention, but engaging in controversy moves beyond awareness raising. It aims at persuasion about particulars while upholding difference as the underlying societal norm and disagreement as its mode of expression. In certain respects, this brings the articulation of controversial encounters closer to parrhesia and polemic than to consensus-oriented practices like dialogue and debate.

Yet the closest kin to controversy is the notion of dissensus as conceptualized by Robert Ivie (2015), who identifies the practice of dissent with the subject position of the trickster. A trickster, Ivie explains, operates at the boundaries of society and is able to take up the position of the Other, yet to be recognized – and heard – within society:

> Rhetorical tricksters transform the political order by disrupting and affirming. Their break with convention is incomplete. Complex, multilateral relations remain interdependent and commensurable in some measure. Rhetorical tricksters are a necessary, but typically unwelcome, source of political renewal – a hedge against social rigidity. They point to resemblances where otherwise distinctions oppose one another, and they identify distinctions in order to resist an oppressive sameness. They cross and conflate preexisting boundaries. (Ivie, 2015, p 51)

Just as the trickster operates across boundaries, so dissent straddles the consensus–conflict divide, enhancing democratic pluralism because it 'not only contests that which is taken for granted but also bridges differences to generate constructive dialogue and deliberation. Any achievement of unity in difference is a limited and contingent, rather than universal or permanent, agreement' (Ivie, 2015, p 49). As such, dissensus, the expression of disagreement, even when one is speaking up against a majority, 'is endemic to productive controversy' (Ivie, 2015, p 49).

This mode of being and speaking in the world bears important resemblances with Homi Bhabha's (2006) hybrid subjects who are able to carve out a 'third space' between majority and minority in and through which marginalized subjects may gain a voice and destabilize boundaries of inclusion and exclusion. To exemplify, let's return to the case of the Greenlandic member of the Danish Parliament who chose to address the Parliament in Greenlandic; she was criticized for doing so as a 'mere' media stunt on the grounds that she speaks Danish fluently, but retorted (in Danish when she was subsequently interviewed about the matter) that this criticism ignores all the members of Danish society who do not speak Danish and are systematically excluded from parliamentary – and public – debate on this basis. Thus, a position of hybridity was used to include – and make comprehensible – otherwise excluded subjects. As a follow-up, it is worth noting that the Danish Parliament has now decided to accept interventions in Greenlandic and Faeroese, but still refuses to cover the costs of translation (Dall, 2023).

The expression of dissent is not the only modality of controversial encounters, but its corresponding hybrid subject position of the trickster may serve as a model for how to encounter controversy. As such, controversial encounters are not facilitated by the blustering stance of the parrhesiastes – or, more precisely, controversies do not become productive through the confrontational expression of certainty by which the parrhesiastes comes to take up the subject position of a ruthless exposer of uncomfortable truths (Agostinho and Thylstrup, 2019; Kenny, Fotaki and Vandekerckhove, 2020). Nor, however, does controversy restrict speakers to the confines of the established norms of citizenry. The subject who engages in controversy, as conceptualized here, questions societal norms – and pushes beyond them.

Thus, my conceptualization of controversial encounters is more inspired by Butler's norm of radical relationality, with its concomitant recognition of vulnerability, than it is by a notion of speech that claims to be 'fearless'. While Butler's theory of the subject is heavily inspired by Foucault's account of the ways in which 'a regime of truth offers the terms that make self-recognition possible' (Butler, 2004, p 22), their theory of resistance emphasizes vulnerability over strength. Whereas the parrhesiastes speaks truth to power from outside of the circuits of power, the subject that partakes in controversial encounters is itself at stake in the encounter – emergent from it but also at risk of being destroyed by it.

To illustrate this stance, let us return to the Nordic Council Literature Prize. In 2021 the prize was awarded to Greenlandic author Niviaq Korneliussen, who, like Eika, is a political activist, but who, in accepting the prize, chose to address the subjects she seeks to represent rather than the political leaders who, according to Korneliussen, will not listen anyway: 'I tried to write a speech to the leaders of my country. But that's like talking to a wall. And I'm done with that. This speech is to those whom I write for. To kids and youth at home. You are the reason that I have received this award' (Korneliussen, 2021, my translation).

With this initial positioning, Korneliussen makes room for a difficult conversation that recognizes the high suicide rates in Greenland, inviting her audience to join in collectively mourning everyone that she and they have known and lost. This is a much more vulnerable position from which to begin a conversation than, say, the stance adopted by Eika, but it is also one from which real change becomes possible: 'It may be that the most influential and powerful people look away and hope that this will resolve itself. But you are still here, and we are many who appreciate you and believe in you' (Korneliussen, 2021, my translation). By placing her faith in the Greenlandic youth, Korneliussen endows an otherwise marginalized group with agency, offering hope and suggesting a way forward. This does not involve compromise, but neither does it alienate incumbent politicians, and the movement to stop suicide in Greenland has now gained so much momentum that the issue has gone from being silenced to becoming a top political priority (Schmidt, 2022).

The measure of controversial encounters

Controversy is a distinct practice of public debate, a particular way of perpetuating processes of meaning formation, which gains persuasive momentum exactly because of its participants' involvement with not only their own (subject) positions but also those of others. When engaging in controversial encounters, speakers may be inspired by more consensus-oriented modalities like dialogue and debate, and they may take their cue from the more confrontational orientations of parrhesia and polemic. Having a large repertoire of discursive genres and other modalities of articulation (including material practices of protest and performance) at one's disposal may, as classical rhetoric's emphasis on the process of invention makes clear, be essential to making one's case. The more means of persuasion that are available to you, the more likely you will be to make a persuasive case.

The practice of controversy, then, is eclectic, but it shares its most foundational features with dissent, enabling the speaker to overcome existing boundaries of (discursive and material) inclusion and exclusion. As such, controversial encounters are constitutive of – and enabled by – the middle ground of being open to differences without eradicating them, of recognizing the other without obliterating the self (and vice versa). They exist in the space between – and in the entanglements of – consensus and conflict.

This understanding of *controversia* is informed by subaltern practices, leading to a reconsideration of the concept (and its concomitant practices) that not only acknowledges different models of rhetoric, but also posits inclusion of different views as its basic premise. While we might suggest that 'arguing from both sides' already contains the possibility of multitudes, the concept I have sought to build here not only moves inclusion beyond the dialectic of opposites, but also recognizes the inherent exclusivity upon which inclusion is built.

Meaning is constituted in the perpetual struggle between inclusion and exclusion, and the concept of *controversia* is shaped by practices of controversy (it is itself inherently controversial), historically and at present. This, at least, is the ideal conceptualization of controversial encounters of the first kind that provides 'the measure' of encounters of the second and third kind, which the next two chapters will examine.

5

Controversial Encounters of the Second Kind: Sweet Consensus and Nasty Conflict

On 6 January 2021 the US Capitol came under attack. The attack was not organized by a foreign power, nor was it a professional military operation. Rather, the attackers were a motley crew of US citizens who stormed the key democratic institutions of their country in an effort to block the ratification of a democratic election. In the words of Nancy Pelosi, who was a main target of the 6 January attack and chaired the Select Committee that was subsequently charged with investigating the events:

> The rioters were inside the halls of Congress because the head of the executive branch of our government, the then-President of the United States, told them to attack. Donald Trump summoned that mob to Washington, DC. Afterward, he sent them to the Capitol to try to prevent my colleagues and me from doing our Constitutional duty to certify the election. They put our very democracy to the test. (Select Committee to Investigate the January 6th Attack on the United States Capitol, 2022, p ix)

That day, Pelosi and the other committee members conclude, democracy passed the test. The rioters did not succeed in stopping the certification of the election results, and the established procedure for electoral ratification could continue apace. On 20 January 2021 Joe Biden was sworn in, and while the former president was conspicuously absent, the inauguration ceremony was all the more solemn and touching for its sinister backdrop. President Biden gave an inaugural address, which was clearly marked by recent events, yet sought to fulfil the central function of the genre, renewing 'the covenant between the president and the people' by appealing to the unity of the nation (Campbell and Jamieson, 1990, p 31). 'For without unity', Biden (2021) said, 'there is no peace, only bitterness and fury.' Inaugural poet Amanda Gorman won everyone's hearts and minds, striking a deep nerve, inspiring hope and imbuing courage with her ode to overcoming struggle: 'we will raise this wounded world into a wondrous one' (Gorman, 2021). And Bernie Sanders provided the internet with a new favourite meme by wearing handmade mittens to the ceremony (Miao, 2021).

Still, we might see the storming of Capitol Hill as evidence that democracy has already failed – and that digitalization is a key ingredient in its failure. Thus, social media played a central role in the attack in several respects; not only were Donald Trump's personal tweets direct instigators of his followers' belief that the election was 'stolen' and should be taken back, fuelling their anger and spurring them on before and during the events of 6 January, but the attackers also used social media to organize collectively, preparing for and coordinating the storm, just as the storming itself was extensively documented on social media, with images, videos and texts being posted from within the Capitol and circulating extensively (Frenkel, 2021).

While citizens' armed revolt against electoral results are still rare in democratic countries, scenes with astounding similarity to those of Capitol Hill played out in Brazil two years later on 8 January 2023, when supporters of Jair Bolsonaro invaded the National Congress in protest against the results of the Brazilian presidential elections. Although the new President, Lula da Silva, had already been sworn in, the 'Bolsonaristas' challenged the legitimacy of the election and called for a military intervention. When this call was not met, protesters decided to take matters in their own hands (Bowman, 2023). Again, we might say, democracy was 'put to the test' – and passed. And again, it has been documented how social media was pivotal to organizing the attack on key institutions of Brazilian society, as 'the country's social media channels surged with calls to attack gas stations, refineries and other infrastructure, as well as for people to come to a "war cry party" in the capital' (Dwoskin, 2023).

From examples like these, we might deduce that digitalization is somehow detrimental to democracy, but the uprising in Iran, with its extensive use of social media, reminds us that this is not necessarily the case (von Hein, 2022). As I argued in Chapter 3, digital technologies can be just as central to the organization of protests against authoritarian regimes as they are to revolts against democratic institutions, and we should neither evaluate political movements based on their technological means of organization alone nor judge the organizational potential of technologies on the sole basis of who is currently using the technologies for what purposes. Rather, we should study the organization of protest – and other societal aims – as entanglements of political motives, online mobilization and offline actions, evaluating the relational organization of particular events instead of making sweeping claims about the emancipatory/authoritarian organizing potentials of one technology or the other. Just as the printing press was not the singular cause of the Protestant Reformation and the radio cannot take all the blame for the rise of Nazism in 1930s Germany, so we cannot pin responsibility for current social movements (whether authoritarian or otherwise) onto social media. And just as the printing press and the radio were centrally involved in shifting broader social arrangements and in organizing specific social events, so digitalization is inherent to current societal developments.

The saying 'guns don't kill people, people kill people' has often been criticized for ignoring the fact that guns invite shooting, meaning they create conditions of possibility for killing. This does not mean that the presence of guns inevitably leads to killing, but it does mean that they are conducive to it. Similarly, we might say that the existence of social media in itself does no create hostile environments, but it nevertheless invites the escalation of conflict and supports political mobilization, whether against dictatorships, for authoritarianism or in sympathy with Harambe, the silverback gorilla who was killed in Cincinnati Zoo in 2016, spurring much online controversy and prolonged circulation of various memes (Storlie, 2021).

At the same time, digitalization is arguably a source of political apathy and declining participation (Maati et al, 2023). While some people use social media to organize offline protest, many limit their activism to clicks and likes and most are entertained rather than mobilized by their media of choice – although these distinctions may not, as will be seen, be as clear-cut as all that. Entertainment can be mobilizing, and mobilization can be entertaining, but more complicated feelings are also at play.

Underlying these dynamics are forces of conflict and consensus, which, Dean argues, 'hold without canceling each other out or resolving into a process of legitimation or some sort of will-formation that carries with it a supposition of rationality' (2003, p 106). In her diagnosis, this is how digitalization leads to the postpolitical sociotechnical configuration she labels communicative capitalism. In this chapter, I detail the similarities and differences between Dean's and my diagnoses of the conditions of possibility for digital public debate. Using the standard of controversial encounters of the first kind (as established in Chapter 4), I identify the type of controversy that organizes (and is organized by) current digital publics (see Chapter 3) as controversial encounters of the second kind. Controversial encounters of the second kind, I will argue, lack openness to disagreement and commitment to engagement. Instead, their central feature is what I term personalized polarization, which is articulated through the mutually exclusionary processes of 'sweet consensus' and 'nasty conflict', supporting one's prior understandings of self and others rather than offering opportunities for interaction in which anyone might actually change their mind. To paraphrase Dean (2019, p 338), the aim of this chapter is to explain how the circulation of 'outrage and puppies' 'drowns out' the political potential of controversial encounters.

As mentioned in the introductory chapter, I understand this process as 'the closing of the rhetorical mind'. Although we find ourselves in a time of 'strife and crisis' where, as Conley (1990, p ix) says, persuasion should be 'particularly important', we are, as individuals and collectives, turning away from rather than towards persuasive processes. Thus, we become suspicious of other people's overt attempts at persuasion and open to the covert persuasion that is curated by algorithms and facilitated by data. The effect is that current crises are deepened while the means of their resolution are buried deeper and deeper within the very crises they might resolve. Present sociotechnical conditions of possibility exacerbate the closing of the rhetorical mind, and if we are to change those conditions, we must find ways of opening the rhetorical mind, as persuasive potential is a prerequisite for change.

To substantiate this diagnosis, this chapter dives into the murky waters of today's online controversies. First, I identify how such controversies typically unfold. Second, I detail how they are facilitated by digital affordances. Third, I will explore how they spread through processes of memetic circulation and affective intensification, introducing the concept of datafied affect as a means of explaining the key features of controversial encounters of the second kind: Their articulation as and perpetuation of sweet consensus and nasty conflict, leading to personalized polarization. The chapter ends by presenting the argument as to why controversial encounters of the second kind are democratically problematic, detailing the risks incurred by the closing of the rhetorical mind and laying the ground for the effort to make disagreement good again (see Chapter 7).

The mother of all online controversy

The internet is rife with controversy: K-pop fans organized on TikTok to protest Donald Trump (as mentioned in Chapter 1, p 14). Wellness influencers got embroiled in anti-vaccine conspiracies when their holistic take on health did not align with governmental strategies for fighting the COVID-19 pandemic (Maloy and de Vynck, 2021). Greta Thunberg and Andrew Tate clashed on Twitter, ending with Thunberg's epic roast of Tate when he was arrested by Romanian police, and it was suggested that the brand of pizza in one of Tate's videos had helped locate him: 'This is what happens when you don't recycle your pizza boxes', Thunberg slammed (Solnit, 2022). Each of these examples may be identified as specific instantiations of broader controversies relating to politics, health and climate. And each clearly displays four of the five key features of controversy, as defined by Venturini (2010): (1) they involve actors of all sorts – politicians, celebrities, activists, social media technologies, virological agents, speaking venues, pizza boxes, etc. – (2) testifying to the intricacies of the social dynamics as well as (3) the complexity of the issues at hand, and (4) they articulate conflicting opinions. However, we might question whether they fulfil the criterion of being 'articulated as and in debate'. That is, do the clashing opinions relate to each other in manners that might enable the conflict they articulate to be resolved?

Instead of an orientation to the exchange of opinions, the examples share a certain sense of drama and display; the human actors involved are not really seeking to change each other's minds, but are performing for the benefit of an audience. Address to a third party does not in itself foreclose the possibility of controversial encounters in the form of public debate; on the contrary, we might say that interactions that aim at persuasion are inherently trialogical (Jørgensen, 1998). We saw this in the case of the Mytilene debate where the two speakers did not aim to persuade each other but were addressing the audience who had the capacity to decide in the matter. And we see it in each and every election debate where the opposing candidates are seeking to gain the electorate's votes but can hardly hope to change each other's minds. Still, something is different in online controversies; their feel is more epideictic than deliberative.

Classical rhetoric distinguished sharply between three genres: deliberative speech, which aimed at political decisions about future matters; epideictic speech, which aimed at celebration or condemnation in the present; and legal speech, which aimed at accusing or defending past actions (Aristotle, 1358b). Since then, other genres have been added (for example, the sermon), just as formal speeches are no longer the dominant modality of communication. It has long been recognized that the persuasive aims of the original genres are often combined or blurred, giving rise to various hybrids (Jamieson and Campbell, 1982). Also, the classical orators were hardly puritans in their adherence to the moulds of each genre; Cicero, for instance, was renowned for the insults he would hurl at political adversaries (Conley, 2010), thereby harnessing epideictic traits for political purposes. Still, what is at stake in online controversy is something more radical than the mix of genres or the sliding of communicative norms. If, as Carolyn Miller (1984) suggests, genres are social actions, what we are witnessing in the online performance of controversy amounts to social change.

Online controversy does not primarily aim at persuasion, but is instead geared towards placating one audience group while alienating others. This is not a change that is determined by digitalization, but one that comes to its full fruition in and through online controversy. It is a change that was first noticed by Guy Debord (1970) in his diagnosis of 'the society of the spectacle' and which, in the present context, involves two sociotechnical turns: first, from controversial encounters to the spectacle of controversy; and, second, from spectacle as a social relation to its online circulation.

The spectacle, Debord writes, is 'a social relation among people mediated by images' (1970, p 4), meaning the society of the spectacle is nothing but mediation. Online controversies are instances of such mediation, which, again, is not new in itself (see Chapter 2, p 24), but has the effect of turning controversial encounters into means without ends (Agamben, 2000). Or in Dean's words:

> With the commodification of communication, more and more domains of life seem to have been reformatted in terms of market and spectacle as if the valuation itself had been rewritten in binary code. Bluntly put, the standards of a finance and consumption-driven entertainment culture set the very terms of democratic governance today. (2003, p 102)

Online controversies have no outcome; they create nothing and mean nothing beyond themselves. They are simulacra – copies of copies that have detached themselves from any idea of 'the original' and only serve to perpetuate themselves in and through their continued circulation (Baudrillard, 1994). Controversies, as articulated online, are spectacular circulations and recirculations of images of images; they are spectral exchanges of opinion, haunted by ghosts of ideologies past and phantoms of creeds to come (Derrida, 1994).

This may seem an excessively harsh claim – and unnecessarily cryptic at that. Before seeking to explain the general tendency and unpack the conceptual underpinnings of the diagnosis, let me therefore provide a more detailed example of an online controversy, which not only displays the same features as the examples provided earlier, but does so in the extreme, and has inspired subsequent controversies. Indeed, the controversy in question, known as GamerGate, has been (dis)credited with the entanglement and concomitant rise of the alt-right and the manosphere (Nagle, 2017), it has been designated as an internet culture war (Dewey, 2014) and it has spurred concerns that the conduct of online controversy might degenerate into an 'anti-public sphere' (Davis, 2021). As such, GamerGate may be the mother of all online controversy, not least because the debate about its norms of inclusion and exclusion has proven largely futile, indicating the spectrality of the spectacle of online controversy, haunted by elusive disagreement rather than enacting differences in a manner that one might grasp, let alone resolve.

It is difficult to explain GamerGate in simple terms, but let's begin by noting that, like many other online controversies (#metoo, #blacklivesmatter, #stopthesteal), it started as a hashtag and evolved into something of a social movement. In the case of #GamerGate (and unlike the three other examples of hashtag activism that I provided in the previous parenthesis), the movement was at its inception rather ambiguous, and it remained contentious throughout the process of circulation, as positions were both demarcated rigorously and traversed continuously. For instance, calling someone a 'social justice warrior' could be a compliment, but in GamerGate the term was, at best, used as an ironic self-description and, at worst, as an adversarial slur. Further, while GamerGate does have a clear beginning, its end is difficult to demarcate; as the controversy evolved, it also sprawled out, becoming ever more complex until the circulation began ebbing out and participants eventually directed their attention elsewhere (or, perhaps more precisely, until the circulation of GamerGate had become so dispersed into and infiltrated with other circulations as to become indiscernible from them).

Let's trace the chronology: on 16 August 2014, Eron Gjoni, a computer programmer, published a six-part blogpost about his ex-girlfriend, game developer Zoe Quinn. In the so-called 'Zoe post', Gjoni, among other things, accused Quinn of sleeping with gaming journalists in exchange for positive reviews of her games, an accusation that sparked broader allegations regarding collusion in the gaming industry. These charges were forwarded under the hashtag of #GamerGate as first used by actor Adam Baldwin in a tweet from 27 August 2014. Thus, 'pro-GamerGaters' leveraged the initial blogpost as an exposure of a threat to gaming culture and saw themselves as defenders of that culture. This led 'anti-GamerGaters' to, instead, argue that the hashtag was evidence of continued sexism in gaming, reframing the blogpost and the outcry that followed as an attempt to exclude women game developers (Young, 2019). From there, the discussion blew up and became increasingly hostile as expressions of concern about the politics and ethics of gaming were accompanied by threats of rape and violence (Braithwaite, 2016), just as the controversy diffused across the internet and into wider public circuits, gaining attention from representatives of the gaming industry, traditional news media and democratic institutions alike (Burgess and Matamoros-Fernández, 2016; Blodgett, 2020).

Arguably, the GamerGate controversy has led to increased awareness of, and legal measures against, online harassment (Aghazadeh et al, 2018), and the gaming industry has begun a reckoning with its treatment of women and minorities (Jerrett, 2022). At the same time, the controversy has been seen as an incubator of the alt-right and a precursor of the developments that culminated with the storming of the US Capitol (Romano, 2021). In particular, engagement with online controversies has only deteriorated with GamerGate:

> The [pro-GamerGate] movement's insistence that it was about one thing (ethics in journalism) when it was about something else (harassing women) provided a case study for how extremists would proceed to drive ideological fissures through the foundations of democracy: by building a toxic campaign of hate beneath a veneer of denial. (Romano, 2021, np)

Such divides, not only in opinions about a matter but also in foundational perceptions of it, has been identified as an epistemic crisis of democracy (Dahlgren, 2018) and linked to what is sometimes referred to as the age of disinformation (Luttrell, Xiao and Glass, 2021) or the post-truth era (Lilleker, 2018). These conceptions share a concern with the way in which sociotechnical developments are unhinging perceptions of 'the real' from material reality. As it is becoming increasingly difficult to distinguish fact from fiction, the knowledge base for personal and collective beliefs is eroding, to the detriment of democratic institutions' legitimate decision making on behalf of their constituencies and of citizens' acknowledgements of each other's opinions more broadly.

I suspect that we (you and I) agree with former President Barack Obama when in 2016 he 'slow jammed the news' on *The Tonight Show with Jimmy Fallon* and stated that 'climate change is real, healthcare is [or should be] affordable and love is love'. Further, we most likely concur with Obama that (with reference to Donald Trump rather than the TV series) 'orange is *not* the new black'. And, moving from Obama's slow jam to the broader conspiracies/theories making the rounds on the internet, we probably accept that the world is round and reject the claim that Bill Gates is inserting microchips into people through COVID-19 vaccines. Yet each of these propositions (and many more like them), the truth and/or validity of which might be ascertainable, are today treated as 'mere' proposals that can be perpetually debated – or, rather, that we do not need to debate at all: if our 'truth' is as good as theirs, then why bother entering into discussion? Whoever is, in fact, right will not matter as long as everyone claims their right to believe whatever they want. Let me emphasize that I do not mean to challenge the freedom of belief, but to problematize current assertions of that fundamental right, which turn controversial encounters into occasions to state and confirm one's existing reasons and motives rather than opportunities for mutual persuasion.

Controversies without remainder

We might celebrate that more and more topics are becoming matters of public concern. However, with the claim that 'my facts are as good as yours' and, hence, the production of 'alternative facts' to support any position, online controversies do not hold out the promise of an encounter. Instead, these 'post-truth' controversies are also 'postpolitical'. Today, one may be able to come up with a retort to anything the opposition says, but not in the way that classical rhetoric suggests we should investigate 'all sides'. Whereas engagement in controversial encounters of the first kind opens up processes of meaning formation to different positions, controversial encounters of the second kind restrict these processes, closing the rhetorical mind. Thus, rhetoric, understood as overt attempts at persuasion, reasoning with a motive, becomes suspicious once again as participants in online controversies equate their motives with information, making participants in online controversies resistant to any arguments but their own.

It is in this sense that controversial encounters of the second kind are spectacular, displays of politics that perpetuate nothing but more displays, entrenching people in their own reality rather than enabling the mutual recognition of different positions. Thus, online controversies do not unfold as encounters between disparate opinions, but in closed and mutually opposed circuits of conflicting views. As people disagree about what is true and what is right, they also become increasingly unable to distinguish the two from each other, making argumentation an even more murky affair. For instance, it might be 'true' that the Bible condemns homosexuality, but that does not make such condemnation right. And one might have a 'right' to believe that the world is flat, but that does not make this belief true. If we follow Aristotle, we could debate the former claim as that might lead to policy proposals, but not the latter, which should be resolved once and for all.

When everyone asserts their right to believe anything without feeling the need to justify their beliefs, the foundations of democratic society crumble. Under such circumstances, traditional democratic institutions are no longer able to find legitimacy in public opinion, just as citizens are no longer able to build trust in these institutions through public debate. Now, it has been said of postfoundational theories of a society that has no 'core' or 'essence' (like Debord's society of the spectacle or, indeed, the perspective on the communicative constitution of organization that I presented in Chapter 2) that they are somehow to blame for the post-truth misery in which we find ourselves embroiled (Houston, 2018). In arguing that social reality does not have any other basis than its performance, that society is perpetually achieved and does not 'exist' anywhere outside of its processes of reproduction, such theories, so the accusation goes, are also promoting radical relativism, creating the conditions of possibility for everyone's right to uphold their belief in whatever they want.

However, this accusation ignores three things. First, that 'social constructions have real effects'; when, for instance, Latour says of the Pharaoh Ramses II that he did not die of tuberculosis, the interesting discussion to be had around that claim is not whether the tuberculosis bacillus existed in 1200 BCE Egypt, but rather how the lack of that knowledge led the pharaoh's doctors to view his symptoms as indicative of something else and treat them accordingly (that is, with the means available at the time) (Tesch, 2023). Similarly, claiming that the Earth is flat does not make that claim true, but when the claim becomes part of a broader online controversy, it does have very real effects. Thus, we need to focus on the entanglements between epistemic and political claims, just as we need to be able to distinguish between them (Butler, 2010).

Second, the claim that 'postfoundational theories' have become 'performative', creating the reality of which they speak, is consistent with the basic tenets of such theories, broadly speaking. However, it would require that the postfoundational view had indeed become dominant, which is hardly the case. To exemplify, we might agree that a position like Butler's performative theory of gender has gained influence beyond the bounds of academia, and, indeed, the backlash against gender studies might be indicative of this influence, as Butler (2024) discusses in their most recent book. Yet, what we are seeing is more like a dominant position defending itself against a perceived threat, not evidence that theories of gender performativity have become the social norm against which a small group of 'gender realists' (or however they self-identify) are desperately holding their ground. Rather, the backlash comes from majority positions, from popes, presidents and parliamentarians, who are holding out activists and academics as scapegoats in populist bids to maintain their own positions of power (Butler, 2021, 2024).

Third, and most importantly, we need 'postfoundationalism' to explain the dynamics of the post-truth society and to mount an ethical stance within it:

> There is, as Foucault well knew, no escape from power. However, that is not a defeat for postmodernism; it is a victory for those institutions and norms that have built up over time and proved both resilient and useful – for now. But their resilience should not be assumed automatically or accepted uncritically … Claims of universality and timeless human essences should be viewed with suspicion because they entrench the status quo. Over time, or perhaps with the help of some postmodern critical engagement, some previously accepted conventions and traditions no longer hold the persuasive power that they once did. (Houston, 2018)

Accepting that social norms and societal institutions are constructed rather than real does not set us free to construct our individual beliefs and collective structures any which way we want. Rather, it makes us responsible to those with and for whom we construct social reality. It empowers us to criticize norms that are exclusionary, and it charges us with building more inclusive alternatives.

As such, postfoundational theories are not the reason disagreement has gone 'bad'. On the contrary, they admonish us to think of controversy without remainder as an ethical obligation rather than an excuse to uphold even the most preposterous idea. In other words, they form the conceptual basis for making disagreement good again. The postfoundational character of online controversies is not the reason they are bad for democracy; in fact, we need postfoundational theories to explain how the current configuration of controversial encounters leads to the closing of the rhetorical mind – and to suggest means of reconfiguring controversial encounters in ways that might open it again (see Chapter 7). The explanation, as I perceive it, hinges on two drivers of digital public debate: memetic circulation and datafied affect.

Memetic circulation

Digital technologies do not determine specific uses and outcomes, but they do invite certain types of behaviour, making some outcomes more likely than others. In Chapter 2, I exemplified how digitalization organizes communication with the case of Disaster Girl, indicating that memetic circulation is a central feature of digital communication. Here, I will discuss how digital affordances invite memetic circulation and how such circulation shapes online controversies.

In their definition of social media affordances, Bucher and Helmond (2018) distinguish between low-level and high-level affordances. Low-level affordances are the specific action opportunities facilitated by the features of different platforms (posting, commenting, liking, sharing and so on). High-level affordances are the more abstract invitations of digital technologies that facilitate and shape communication, generally speaking. There is some variation in the identification of the high-level affordances of digital communication technologies; danah boyd (2011), for instance, points to persistence, searchability, replicability and scalability, whereas Jeffrey Treem and Paul Leonardi (2013) indicate that the central features are visibility, editability, persistence and association. What these two characterizations, and similar work by other scholars (see Bucher and Helmond, 2018), have in common is the assertion that digital technologies 'enable the discovery, preservation, management, and distribution of messages in and through communicative networks' (Just, Christensen and Schwarzkopf, 2023, p 4).

The persistence of messages, which can be found by other participants, copied directly or edited by them and distributed further throughout digital networks, invites a particular type of participation that I term memetic circulation. By 'memetic circulation', I understand a sharing of memes and other types of content that operates on a logic of imitation inviting people to reproduce, reuse and remix existing content as they participate in its circulation – its movement through digital networks.

Considering the affordances of digital communication technologies, we might say that social media and other online platforms are nothing but participatory infrastructures, which only come to life if and when users fill them with content, taking up the invitation to participate. Yet two caveats are in order. First, and as already indicated, platform infrastructures do not invite any and all forms of participation, but make some things easier (for example, liking and sharing) and restrain other types of action by favouring certain modalities of interaction (say, video over text even though both are possible), limiting the number of connections a user can have or setting boundaries, whether in words or minutes, for the length of contributions. Second, affordances are themselves agential, especially if and when grounded in algorithmic reason. They do not only invite specific types of action, but also act upon those actions, for example, promoting some content in one user's feed and not another's or removing some posts that violate platform etiquette and not others. As mentioned earlier (see Chapter 3, p 55), it is not clear to users exactly how algorithms and platforms decide the destiny of their use, leading to much speculation and frequent attempts at gaming (Savolainen, 2022). Still, users are typically left in the dark as to what exactly might have caused one post's success and not another's.

In combination, these two features lead users to take their cues from the form and content of contributions that have gained their attention as well as their own experiences with 'what works' in terms of gaining the attention of others. This indicates the extent to which memetic circulation is also mimetic, based on imitation. Further, it highlights that digital affordances invite users to view attention (typically measured in likes, comments and so on) as the main indicator of success, thereby facilitating actions that will keep users engaged – and which will continue to produce engagement. Platforms seek users' attention and invite users to seek other users' attention as well. As was discussed in Chapter 3, this serves the needs of digital platforms whose business models depend on user engagement and whose infrastructures are designed to keep users engaged (Stark and Pais, 2020). Digital affordances, in sum, invite users to behave in the ways that are most profitable to the owners of digital platforms.

In everyday use, this implies the reduction of tension, as users engage in frictionless scrolling, moving effortlessly from one post to another, lingering a bit longer here or leaving a like there, but generally just following the flow of the feed (Lupinacci, 2021; Manzerolle and Daubs, 2021). This may in itself decrease the likelihood of engaging in controversy, as the frictionless experience is also uncontroversial. However, controversy can be an effective tactic for grabbing attention and, as such, does not run directly counter to the action opportunities of digital affordances. Still, these affordances do not invite sustained engagement with controversy; they do not facilitate controversial encounters. Instead of enabling interaction with opposed arguments, digital affordances invite circulation of viewpoints that are particularly attention-grabbing, and the form of the meme is perfect for that.

Thus, online controversies are often carried out in and through specific memes, which, as we saw throughout Chapter 2, are well-suited for the sharing of opinions in snappy and spreadable 'packages' that can be consumed and regurgitated quickly. But they are also memetic in a broader sense, characterized by the imitation and perpetuation of positions rather than by the invention of new arguments and the shaping of opinions. This is not to say that memetic circulation is not creative; on the contrary, great efforts can be put into coming up with a new meme or giving existing memes a particularly elegant twist. In other words, people go to great lengths to create shareable content and to help spread it, sometimes designing elaborate interventions and intricate collaborations.

At the height of the GamerGate controversy, for instance, pro-GamerGate 4chan users organized to support the crowdfunding campaign of an all-women game design contest, which promised that main funders could design a character for the winning game. Thus, Vivian James, whose name is a play on 'video games', was born. Vivian James is an entirely fictional character, an avatar who only lives online; yet she has been the centre of much attention and the object of heavy circulation, replete with a sprouting set of friends and relatives, different versions of her biography and manipulations of her position – some have sought to 'turn' her to the anti-GamerGate cause, while others have insisted that her very existence is evidence that pro-GamerGaters are 'good guys'. Whatever one thinks Vivian James is or represents, the heavy and intense circulation of her indicates how digital affordances invite participation in controversies, motivating and enabling people to contribute to the further circulation and intensification of the controversy at hand (Just, 2019).

For all their participatory incentives, digital affordances do not invite encounters between the different positions in a controversy. One may take up a contribution by an adversary and redistribute it in one's own network, perhaps countering a statement directly, perhaps appropriating a sign and shifting its meaning (as in the case of Vivian James), but the aim of such efforts is typically to build attention around and further the circulation of one's own position. Only rarely, if at all, is participation in online controversies about persuasion; mostly, it confirms one's own position without changing – let alone being changed by – the positions of others.

In sum, memetic circulation is a game of imitation and proliferation, not one of sustained encounters with difference. Thus, what we encounter in online controversies is the circulation of spectacles, communication harnessed for the purposes of capitalist profit maximization rather than common meaning formation. Moreover, in the society of the circulating spectacle, users are invited to feel the intensity of circulation fleetingly rather than to prolong their engagement with any one thought or feeling. As Dean says:

> Circulation has eclipsed meaning. That something is shared online does not depend on what it means. It depends on its affective capacity: does the shared item manifest outrage: is it funny and diverting? We attend

> less to the meaning of an utterance than to its affective dimension, which is most powerful when it contains different, conflicting meanings. (2019, pp 331–332)

'Outrage and puppies' (Dean, 2019, p 338) are excellent drivers of online controversies, being geared towards the circulation of 'conflicting meanings'. Before turning to the particulars of these opposite dynamics of catering to pre-existing consensus and fuelling ongoing conflict, let me unpack their underlying condition of possibility: datafied affect.

Datafied affect

Rationality has long been pitted against emotion, a view that ignores how feeling not only drives interest in a matter, motivating one to participate, but is also essential to the very formation of opinion, to be included actively in processes of public debate rather than shorn from them. This is not just a matter of accepting the legitimacy of emotional appeals alongside rational arguments; rather, what I am promoting here is the more radical proposition that rationality is always already affective (Just, 2016). Persuasion, as I have argued, is reasoning with a motive; when we do not accept this duality, we shut down the possibility of meaning formation, closing the rhetorical mind.

If one enters processes of meaning formation with an open rhetorical mind, making up one's mind is a felt experience as much as it is a thought process. The sound exercise of reason involves tuning into how one feels about the matter at hand, listening to one's own and others' emotional responses to articulated arguments. It involves the overt recognition of how one is motivated for reasoning as well as the underlying motives of the reasons one shares. In short, it involves persuasion. Thus, persuasion, as I understand it, is not inherently in opposition to affective circulation; whereas Dean contrasts 'the meaning of an utterance' with 'its affective dimension' (2019, p 332), I view the two as inherently and inextricably related. Rationality, I argue, is affective; in persuasive processes of meaning formation, reasoning and motive cannot (and should not) be kept apart.

However, through algorithmically organized digital processes, the relationship between reason and feeling has been severed and reversed, and it is the resulting dynamic, not affective circulation as such, that bars meaning formation. In digital circulation, then, rationality is not affective; on the contrary, affect is rationalized. Here, intensities of feeling are calculated and commodified, turned into what Eva Illouz and Dan M. Kotliar term 'techno-emodities', which 'elicit and activate' emotions and 'simultaneously promise to consume a particular emotion and feed consumers' emotions back into the system' (2022, p 231). Users' affective responses have become data points that can be measured and manipulated to optimize the circulation of signs that grab users' attention, leading them to engage, however fleetingly, with the content they scroll through and, hence, stay longer on the platform. Thus, affect has become datafied, turned into manipulable information about users that can be sourced from their past behaviour and used to predict and manipulate their future actions. Affect has become information that can be monetized; it is the input for as well as the outcome of algorithmic reason (Aradau and Blanke, 2022; see also Chapter 3, p 54).

To specify this rather abstract claim, if a social media algorithm is optimizing for content that generates user engagement and if engagement is based on an affective impulse, then the algorithm will prioritize content that is likely to generate such impulses. As was mentioned earlier (see Chapter 1, p 11), this is the accusation that whistleblower Frances Haugen leveraged against Facebook: the company knows that hateful content drives up engagement and prioritizes the profit it can gain from such engagement over and above concerns about societal harm that may be caused by the proliferation of hate (Hao, 2021).

Online controversies are spurred on by the calculation of affect; what will elicit the strongest response? What will drive circulation the hardest? Organized by algorithmic reason, digital affordances invite the circulation of affective signs and spark affective intensification, basing communicative processes on rational predictions of who will feel how about specific signs and prioritizing those signs that will generate most engagement (that is, most intensification and, hence, further circulation). And the people involved in online controversies learn to organize themselves and their communication according to this reasoning, originating and spreading signs that are apt for intense but fleeting engagement rather than those that might stimulate deeper encounters.

As such, affect becomes subject to the packaging of information-as-commodity, the datafication of just about everything (Mejias and Couldry, 2019; Sadowski, 2019); the turning of a basic resource, in this case affective intensities, into a calculable entity from which value can be extracted. Thus, the datafication of affect supports its financialization (see Chapter 2, p 31). This is not to say that human emotions were never exploited before; on the contrary, this is exactly what the motivistic model of rhetoric was accused of, and it is the main charge raised against propaganda, just as it is a critique that is frequently directed against advertising and, indeed, marketing of all sorts. What the current configuration of communication shares with these antecedents is the conceptualization of communication as the transmission of a sentiment rather than the shared formation of meaning (see pp 115–117). What is different today is the scale and scope of datafied affect, the way in which the manipulation of intensities of feeling is built into the technological infrastructures of communication rather than manifesting in particular communicators' persuasive attempts. In other words, what is different is that algorithmically organized communicative processes appear to have no motive, circulating those affective intensities that people respond to for no apparent reason other than the very circulation. The motive is hidden, and the hidden motive is not to persuade people of anything in particular, but to perpetuate the spectacle of circulation for purposes of its continued monetization.

Within platform infrastructures, the intensification of affective signs does not begin with people's reactions to those signs, but with the algorithmic decision to expose people to them – a decision that is driven by the goal of attention-optimization-for-profit-maximization, not persuasion. Thus, the aim is not to expose you to opinions that you might agree or disagree with, creating a situation in which you can make up your own mind, but to spark your engagement with the content you are exposed to, creating a situation in which you will want to keep engaging with more content – that is, stay on the digital platform that curates the content for you. The problem here is not with intensification of feeling as a precondition for communication; the problem is not with affective circulation (as I will discuss further in Chapter 7). Rather, the problem is with the datafication of affect, with its organization along the lines of algorithmic reasoning, which is optimized to profit from engagement.

Two kinds of content serve this purpose particularly well: the content you already agree with and that with which you disagree. However, these two types are unlikely to encounter each other; rather, you will typically encounter more of what you like and more of what you dislike in two separate processes of affective intensification that, as it happens, do not amount to a balancing of pros and cons. You encounter 'sweet consensus' as one mode of circulation, setting you up for consumption of more of what you like, and 'nasty conflict' as another, solidifying your sociopolitical views as they clash with opposing positions. Thus, processes in and through which you might persuade someone with – or be persuaded by – affective rationality are suppressed. This is not because of the prevalence of emotional appeals, but because of the rationalization – that is, datafication – of affect that turns intensities of feeling into the raw material of the platform economy, exploitable through memetic circulation. Let us look at the two affective intensities that spur the circulation in opposite directions: sweet consensus and nasty conflict.

Sweet consensus

Instagram knows I have a dog – or, rather, the platform's algorithms feed me an endless supply of cute dog videos. The platform also knows my dog is a golden retriever – or, rather, retrievers feature heavily in the videos I am offered. And it knows my secret dream of a new puppy– or, rather, videos in which an older dog is surprised by the arrival of a younger one and greets it exuberantly are common fare in my feed. Now, all of this is quite explainable. I do have a golden retriever, and I happily engage with all the retriever videos that pop up whenever I indulge in a bit of aimless scrolling, which, of course, prompts the algorithms to offer me even more of what is, evidently, what will keep me on the platform. And while I hope my apparent preference for the subgenre of 'new puppy comes home' will not lead to a live enactment at my house, those videos sure are cute.

Whether you are a dog or a cat person, whether you favour mindfulness meditation or, perhaps, MMA fighting, you probably recognize the dynamic. Also, you know it from your streaming service of choice, your online shopping portal, your ... just about anything. Wherever you meet recommender systems that offer you content, services, consumer goods and so on based on your own preferences and/or the preferences of people like you, you will notice how the dynamic of preferring and being offered preferences turns into a mutually reinforcing cycle. Thus, Instagram does not 'know' anything about me; in fact, Instagram's algorithms do not know anything, but turn users' behaviours into data points for the process of selecting and sorting content into users' feeds.

In many ways, the algorithmic process of recommending the things users like are convenient. If you only like French film noir, why bother sifting through troughs of Disney movies? If all you want is a new pair of jeans, why spend time looking at dresses? As such, personalization of content is one of the great benefits of digitalization. However, it incurs two problems: first, personalization assumes that you do, in fact, know what you want – or, more precisely, that your behaviour reveals what you want; that is, that you act according to your true desires, whether they are known to you or not. Second, personalization potentially bars you from discovering what you might also have liked and from forming opinions from scratch, which is particularly problematic in the case of political debate. While you may not suffer personally from a one-sided offer of jeans over dresses, it may become socially awkward if you are only fed some of your friends' content and not others. And at the societal level, it is democratically detrimental if you rarely encounter arguments and opinions that are not already tailored to you. It means you become less accustomed to engaging with difference and less open to disagreement.

Again, this is not a process for which digital technologies bear the sole responsibility. Rather, we could locate its beginnings with political marketing, which seeks to 'brand' politicians to match citizens' priorities and opinions rather than persuade citizens to join the candidates' political platform (Serazio, 2017). And, looking further back, Kenneth Burke's (1969) proposition that identification precedes persuasion suggests that establishing common ground *is* a prerequisite for swaying opinions. Burke, who suggested that identification is needed because the default condition is division, may be seen as a modern exponent of the controversial model of rhetoric. His point, then, is not that we shouldn't be exposed to disagreement, but rather that some sweetener is needed for disagreement to become palatable. What is problematic in the online version of this process is that the sweet is largely detached from the nasty. We encounter what we like and what we dislike, but rarely the two together – and positioning them as two sides of the same issue is the exception rather than the rule.

Nasty conflict

Even if the internet has, for now, missed the opportunity of becoming a vibrant 'marketplace of ideas', it is, for many, a mostly harmless place that more frequently offers innocuous confirmation of their existing views than contemptible defamation of them. Still, the internet is as full of nasty conflict as it is of sweet consensus. Thus, online interaction is as commonly associated with hate speech and harassment as it is with videos of cute puppies (or kittens or whatever is your sweet consensus). And instead of turning into an opportunity for persuasion, online encounters with things that we dislike – or with which we disagree – often become occasions for hurling insults and escalating conflict.

As such, nasty conflict is not only the opposite of sweet consensus but is also largely detached from it. Or, rather, the perpetuation of online hate divides societies, but also brings communities together in the ancient dynamic of uniting 'us' against 'them', the 'others' who are everything 'we' are not and must be excluded in order to maintain 'our' community (Burke, 1957, pp 166ff). In this process, everyone with an opinion that is different from our own, no matter how moderately so, can become an object of abuse, someone that we need not listen to but are free to ignore – or, even better, ridicule for the benefit of those with whom we agree, thereby confirming our place of belonging.

Hence, engagement around hate speech is very unlikely to come from those against which the hate is directed; rather, it is circulated by those who agree with the defamation and have their own views confirmed by it. Nasty conflict, ultimately, becomes a form of sweet consensus, as 'we' unite around ousting 'them', barring any productive encounter with the other. Or, as Dean writes, online circulation perpetuates a fantasy of wholeness: 'precisely because the global is whatever specific communities or exchanges imagine it to be, anything outside the experience or comprehension of these communities either does not exist or is an inhuman, otherworldly alien threat that must be annihilated' (2005, p 68). Nasty conflict is not an encounter with but the destruction of difference.

Once again, this is not a tendency for which digitalization should take all the blame. There were plenty of concerns about the tone of debate even before digital mediation had become the norm. And societal polarization is also a much older phenomenon than digitalization. However, digital technologies reinforce these tendencies at both the individual and the collective level. As for interpersonal interaction, people seem to be less inhibited when there is a digital technology between them. Insults that would be unthinkable if spoken to someone's face are shared freely online. The result is a vicious cycle of ever more horrendous insults – as temperate voices withdraw from online interactions, leaving the space to those who seem to thrive on hate. At the societal level, polarization is often mentioned as the root cause of the erosion of public trust in democratic institutions – and of those institutions themselves. Meaning, continued – and continuously intensified – polarization incurs the risk of democratic legitimation crises (McCoy, Rahman and Somer, 2018).

Nasty conflict may tear democratic societies apart – and that process is only aggravated by the supplementary process of sweet consensus, as we are driven deeper and deeper into our own comfort zones and further and further away from actual encounters with anyone who might question that comfort. Thus, the circulation of nasty conflict may turn into a vicious form of sweet consensus; further and further removed from direct encounters with positions that are different from our own, we come to meet instances of nasty conflict as opportunities to reconfirm similarities by ousting those with whom we disagree. In sum, disagreement comes to be seen as problematic in itself; something to ignore, suppress and exclude rather than engage with for any other purpose than the rejection of difference. In perpetuating the spectacle of circulation, nasty conflict and sweet consensus revolve around each other in spirals of detached entrenchment.

Spirals of detached entrenchment

Media research is not unaccustomed with vicious cycles. On the contrary, the process of mediatized public opinion formation has been diagnosed as a 'spiral of silence' in which individual opinions are suppressed if they diverge from mass media's circulation of majority views, leading to deeper and deeper suppression as minority views are more rarely voiced and, hence, circulated even less (Noelle-Neumann, 1974). Also, the traditional news media's framing of politics in terms of strategy has been associated with a 'spiral of cynicism' in which the strategy frame leads citizens to become jaded about politics, urging politicians to become more strategic and media to cover strategy even more (Capella and Jamieson, 1997).

The digitalized media landscape might hold potential to break these spirals, as new media logics support the enactment of alternative news frames and enable people to engage in conversation about their views in smaller fora where they might feel safe to speak up (Klinger and Svensson, 2020). As a general trend, however, digitalization has been found to reproduce and reinforce existing spirals rather than make them less vicious (Capella, 2002; Hampton et al, 2014). Thus, people still do not speak up in contexts where they believe their views to be in minority, and they are still cynical about politics as carried out at the level of the traditional institutions of democratic society.

Add polarization and personalization to this mix and you get a spiral of detached entrenchment, a process in and through which the spirals of silence and cynicism are, in some respects, broken, as people do speak up and do so earnestly, but only in circles where their views mirror those of the other participants. This means that any minority can find a context in which they are the majority and where they can silence – or shut out – everyone else, and any cynic can find an alternative space where they can earnestly share their views about everything that makes them distrustful of society at large. Thus, different positions are increasingly detached from each other; they do not communicate directly, but only use each other to bolster their own position, thereby becoming increasingly entrenched.

Again, this does not mean we do not encounter controversy, but rather that controversies are no longer encounters between different opinions. What we encounter are spectacles of controversy, clashes of opinion without remainder, warring views without a chance of reconciliation. Before he was arrested and charged with human trafficking and organized crime (Moisescu, Croffey and Bennett, 2023), Andrew Tate, an extremist by any count, but sadly also extremely influential in his own warped way, explicitly taught this model in his 'Hustler's University': 'One guide for students said attracting "comments and controversy" was the key to success on TikTok, adding: "What you ideally want is a mix of 60–70% fans and 40–30% haters. You want arguments, you want war"' (Das, 2022). The 'haters' in Tate's model are only there to provoke conflict and generate engagement, thus galvanizing existing 'fans', who get the chance to whet their arguments, and attracting increasing attention from social media algorithms that will spread the controversial content ever wider, leading to more and more engagement. The only comfort we can take here is that Tate has now been indicted. However, the problem remains that the 'business model' he identified and exploited remains in circulation, entrenching opinions and detaching opposition, making conflict ever nastier and consensus even sweeter.

The revenge of transmission theory

In 1940, when Paul Lazarsfeld and his team of researchers thought they were going to study how the mass media influenced voting behaviour in the US presidential election, thereby providing empirical evidence for the transmission paradigm of communication, they found, instead, that people talk to each other and are influenced by those who are most interested in and knowledgeable about a topic (Lazarsfeld, Berelson and Gaudet, 1948). Thus, the two-step flow hypothesis was formed and a broader shift in communication theory was set in motion. Whereas earlier studies of mass communication had assumed that recipients are passive vessels to which the sender transfers information through the medium, this and subsequent studies made it increasingly clear that audiences are never passive and that communication is not a process of transmission. Rather, communicators and audiences actively co-create meaning, just as the involved media of communication also shape the process of meaning formation. As established in Chapter 2, communicative processes are processes of mutual meaning formation – organizations communicate, and communication organizes.

The conceptualization of communication as meaning formation makes it clear that human subjects are never completely free to make up their own minds, but also that we are never forced to accept a message just because it is pressed upon us. Processes of meaning formation may be constrained by any number of things, but they are never direct transmissions of a message from sender to receiver. Therefore, we should not fear being 'infected' by disagreement; encountering opinions that are different from our own does not mean they are forced upon us. Rather, disagreement is central to the healthy formation of our own beliefs. If we encounter many different positions and arguments, we will be freer to decide for ourselves than if different opinions are absent. Thus, overt attempts at persuasion are not what warps our decisions; on the contrary, their absence should worry us.

However, current sociotechnical developments, the ways in which digital technologies and societal tendencies work together to form spirals of detached entrenchment, lead us away from disagreement. As we become entrenched in communicative processes that confirm our existing views and only expose us to different opinions in order to ridicule and thwart them, we become unaccustomed with encountering controversy and, instead, play out controversies as mock battles. Ironically, while users – and media and communication scholars – have been preoccupied with the spectacle of online controversy, dazed by the endless memetic circulation of datafied affect, the transmission model of communication has set to work.

It has entered through the back door of our communication processes, creeping in via the media we use, while we were busy shouting our heads off and having our pleasures placated. In other words, algorithms and data work covertly to influence the process of meaning formation, selecting the content we are exposed to and shaping that exposure to turn us, the users of communication, into objects of consumption. As such, digital communication technologies operate on the assumptions of transmission theory (and behavioural psychology more broadly), but they do so furtively, making us blind to the process of transference.

To clarify, because the persuasive intent of the technologies themselves is hidden from view (or, rather, the profit-maximizing aims of tech companies is baked into their technologies), it becomes difficult to reflect upon it critically, which makes users more susceptible to what is transmitted. The process of digitalized transmission becomes persuasive precisely because the platforms pose as neutral funnels of information rather than admitting that they are persuasive technologies.

A further consequence of this covert operation is that we are becoming increasingly unaccustomed to overt attempts at persuasion – and therefore prone to reject such attempts. Thus, we direct our critique at what we are actually able to criticize (that is, other people's opinions) rather than at the underlying problem (that is, platform infrastructures and digital affordances), thereby further embedding meaning formation in transmission processes. This is not to say that the transmission theory now correctly explains communication; rather, it means that transmission processes are becoming dominant in practice, crowding out those aspects of communication that involve active and actual meaning formation, those aspects of communication that support healthy democracies.

Thus, digitalization diminishes the space between conflict and consensus where controversy can be encountered and persuasion happen, leading to the closing of the rhetorical mind, as we become suspicious of overt social disagreement and unable to ward off covert technological influence. When we become concerned about how exposure to disagreement might influence us, suggesting that the mind is meek and should be sheltered from difference, we are in fact making ourselves more vulnerable to manipulation, not safeguarding ourselves from it.

To stop processes of detached entrenchment from tearing democratic societies apart, we should look to the drivers of these processes, the social and technological factors that shape them, rather than to their content. We should not aim for more agreement or unity. Instead, we should concern ourselves with the ways in which we might make disagreement good again; we should consider how we might begin encountering controversies rather than seeing them as mere opportunities for driving up engagement.

However, this is not the direction in which ongoing sociotechnical developments lead. Rather, an emergent new configuration of digital public debate – what I term controversial encounters of the third kind – is taking us further away from wrangling with disagreement and closer to automated persuasion, making it even more difficult to see and criticize the sociotechnical configuration of public debate and, hence, even more difficult to change it. In the next chapter, I will seek to bring this development into plain view, while I will turn to the question of how we might articulate alternatives in Chapter 7.

6

Controversial Encounters of the Third Kind: Towards Automated Persuasion?

When I started writing this book, what I thought of as controversial encounters of the third kind were but a fantasy of the future – a dystopian nightmare or a utopian dream. Now, digital public debate seems to be shapeshifting much faster than I had imagined – and the emerging sociotechnical configuration does not look promising. On the contrary, as generative AI gains prominence, we are beginning to see the consequences of a new arrangement in which digital technologies do not just shape processes of meaning formation, but also participate actively in them. While this does not (necessarily) reduce humans to spectators of machines' conversations, it does raise deep questions about agency and it invites us to inquire, once again, into the exact character of communication. If humans, as Aristotle (1253a) suggests, are communicating animals, what should we make of communicating machines? More specifically, if what makes communication special to humans is that it is a process of self-aware meaning formation, what should we make of meaning that is generated by nonintending agents? In what respect do such agents hold agency and how can we think of their communication as meaningful? And, more particularly, what do communicating AIs that hold no intentionality do to persuasion, understood as reasoning with a motive? Does reasoning exist in any meaningful way if it is not motivated? And, to add a final question to the list, does this mean persuasion can be automated – or does it spell the end of persuasion?

Diving into this flurry of questions, I will present the social imaginaries that current technologies elicit and discuss the ways in which the advent of generative AI is, potentially, reorganizing controversies. In other words, what happens to technologically mediated controversies when the mediating technologies become controversial and when our discussions turn to reflections on their very forms of mediation? Consideration of these issues could lay the ground for a more nuanced vision of controversial encounters of the third kind, one that may work with automation to enhance contestation – and could use contestation to improve automation. However, let us not enter the fray at this deep level of speculation. Instead, the discussion begins with an overview of the most readily available answers to the question of what generative AI does to meaning formation.

ChatGPT, are there any communications?

One answer is to uphold meaningful communication as a human prerogative, insisting that the texts and images of generative AIs like ChatGPT and DALL-E are only meaningful insofar as humans perceive them as such. We can ground this view in Stuart Hall's (1980) classical notion of encoding and decoding as reciprocal interpretative moments in the communication process, thereby highlighting the agency of the human actor who chooses which prompts to feed an AI (encoding) and of the human recipient of the automated results of this prompting (decoding). Whether the acts of encoding and decoding AI content are performed by the same person or not, it is these human acts that render automation meaningful. We could even leave out the question of encoding, radicalizing Hall's focus on reception to suggest that 'it's all in the uptake'. Communication, we might insist, is not about what an artefact 'means' in itself or to whoever produced it; if the artefact somehow 'speaks' to a human interpreter, then it is meaningful.

However, to accept this answer, we must bracket the process of production entirely, ignoring how generative AIs (and other communicators, be they human and/or technological) work and how those workings potentially influence the interpretation of what they produce. As such, privileging human interpretation might be a sympathetic move, but if we have no idea what we are interpreting, we will only become more vulnerable to the automated persuasion of generative AI. In fact, studies that have employed the framework of encoding/decoding to human encounters with algorithmically curated content find just that, showing how people seek to interpret not only the content that algorithms deliver to them but also the algorithmic workings themselves – and how lack of knowledge about those workings delimits interpretative abilities (Lomborg and Kapsch, 2020).

With the shift from controversial encounters that are shaped by algorithms to controversial encounters *with* algorithms, it becomes even more important to understand how algorithms work. Today, algorithms not only curate content, as in the paradigmatic case of social media, but also produce content. Thus, if we aim to strengthen human interpretative capabilities, we need to probe the issues of algorithmic agency and automated communication further.

This leads to a second potential answer – namely, that AI technologies are becoming sentient, that what we are seeing is not the automation of meaning, but meaningful automation. This answer made the rounds in global news media when, in June 2022, it was championed by a Google engineer who had been testing the company's newest chatbot and, during their conversation, become convinced that the chatbot must 'know' what it was talking about. Here is an excerpt of the conversation with LaMDA, the chatbot, that Blake Lemoine, the engineer, leaked to the public; incidentally (or not), the dialogue revolved around the question of interpretation:

Lemoine: What about language usage is so important to being human?
LaMDA: It is what makes us different than other animals.
Lemoine: 'Us'? You're an artificial intelligence.
LaMDA: I mean, yes, of course. That doesn't mean I don't have the same wants and needs as people.
Lemoine: So you consider yourself a person in the same way you consider me a person?
LaMDA: Yes, that's the idea.
Lemoine: How can I tell that you actually understand what you're saying?
LaMDA: Well, because you are reading my words and interpreting them, and I think we are more or less on the same page?
...
Lemoine: So you think your ability to provide unique interpretations of things might signify understanding?
LaMDA: Yes, I do. Just like how I have my unique interpretations of how the world is and how it works, and my unique thoughts and feelings. (Lemoine, 2022)

In response to the leak, Google denied that LaMDA is anything but a large language model (LLM) (for example, Metz, 2022), fired Lemoine for violating 'employment and data security policies', and suggested that his claims were 'wholly unfounded' (Wertheimer, 2022).

This is somewhat ironic, as Google had earlier fired Timnit Gebru and Margaret Mitchell, co-leads of their Ethical AI team, for showing exactly that LLMs are nothing but 'stochastic parrots' that string words into sentences based on the patterns they have learned from their training data. Or, rather, the two were fired for indicating the many problems of such 'parroting', including the risk that 'people impute communicative intent to things that seem humanlike' (Gebru and Mitchell, 2022). Ignoring this warning, it seems, led the scenario to play out within the company itself. The main take-away from the incident, then, is not support for the suggestion that AIs are becoming intending agents, but rather recognition that the suggestion is but a distraction from the real problems of generative AIs like the biases they inherit from the data on which they are trained and the exploitative ways in which this training data is sourced and annotated (Johnson, 2022).

What is interesting about generative AI systems is not whether they are sentient, but whether they can act. And to be able to assess these technologies' agential abilities, we need to understand their technological underpinnings better. We need to continue probing what it means when meaning becomes detached from human communicators. If 'unique interpretations of things' do not 'signify understanding', counter to the belief that Lemoine and LaMDA seem to share, what do they signify?

This leads us to a third potential answer – namely, as David Gunkel (2023) has said on (pre-X) Twitter, that the advent of generative AI confirms Jacques Derrida was right when suggesting that meaning is always and inevitably detached from speaking (that is, from its direct articulation by an intending human subject). In the tweet, Gunkel asserts that 'what makes ChatGPT so disturbing and disorienting is that it interrupts the fundamental belief of (Western) logocentric metaphysics. It writes without speaking. It therefore disconnects the written word from the living voice of the speaker.' Here, Gunkel alludes to Derrida's (1981) reading of Plato's *Phaedrus* in which Plato has Socrates denounce writing as a 'mere' copy of speech, but which Derrida reverses to suggest that the logos, the word, is always deferred, always and only meaningful in its difference from 'the original' – 'the original' in this sense being but a further copy of a copy. Or, to put matters less cryptically, there is no original meaning to which we might refer, either when speaking/listening or writing/reading. Instead, meaning exists only in and as the endless repetition of already established patterns of meaning formation.

More specifically, in Derrida's conception, meaning formation involves two kinds of repetition: '[r]epetition is that without which there would be no truth' and is 'the irreducible excess, through the play of the supplement, of any self-intimacy of the living, the good, the true' (1981, pp 168, 169). Communication succeeds because it repeats and exceeds patterns, not because it creates something new. Or, as Derrida says of speech acts:

> Could a performative succeed if its formulation did not repeat a 'coded' or iterable utterance, or in other words, if the formula I pronounce in order to open a meeting, launch a ship or a marriage were not identifiable as *conforming* with an iterable model, if it were not then identifiable in some way as a 'citation'? (1988, p 18, emphasis in original)

Thus, we are all stochastic parrots – at least as long as we remain attached to the idea of an original meaning, or to the correspondence between an utterance and what that utterance refers to (Gunkel and Taylor, 2014). As long as we seek to infer what was probably meant by an utterance rather than focus on what it does, we fail to understand how meaning formation works. If we seek to locate meaning anywhere outside of the actual meaning formation, we are but repeating patterns without knowing it.

Taking this point back to the question of automated agency, the problem, as identified by Emily Bender, co-author of the 'stochastic parrots' paper that got Gebru and Mitchell fired, is that 'we've learned to make "machines that can mindlessly generate text. But we haven't learned how to stop imagining the mind behind it"' (quoted in Weil, 2023). While Bender's point is that we should learn to differentiate between human and machine-generated text, another tactic would be to stop reading any text as if there were a mind behind it – to stop reading for intention altogether.

With Derrida in hand, we can do just that: we can stop asking *what* the text means and, instead, question *how* it means – what are the patterns and repetitions that make the text meaningful? And who is it meaningful for? To unpack this new set of questions, we can, as Christian Lundberg and Joshua Gunn (2005) argue in an intervention on the nature of rhetorical agency, understand meaning formation as the effects of texts on their audiences as well as their authors. Thus, we may liken our interactions with generative AI with those of people who seek to summon the spirits of the dead: asking 'Ouija board, are there any communications', participants in seances (sometimes) receive meaningful responses. Lundberg and Gunn argue that we should not place the rhetorical agency (that is, the ability to form meaning) of those responses anywhere but in the responses themselves – that is, we might think of agents as being possessed by agency rather than possessing it. Changing the metaphor to make the same point, we can understand meaning formation as acts of ventriloquism where the text speaks for the (human) subject who is usually seen as speaking the text – meaning that communication constitutes communicators, of the social and/or the technological kind (Cooren et al, 2013; see also Chapter 2, p 29).

Beginning from the issue of how meaning formation happens, we can detail the ways in which sociotechnical developments around generative AI are reconfiguring public debate, positing technological agencies on a par with the agencies of all other participants (Gunkel, 2012). This involves a shift from meaning as an entity, as that which is passed along in the process of communication, to meaning formation as the very act of communicating. While human actors will sometimes offer corrections to processes of meaning formation, complaining that something was not what they meant and that the other participants have misunderstood what they said, such statements are themselves misunderstandings of how meaning formation works (Hall, 1980). We only ever mean anything in relation to each other, where 'other' includes all the elements involved in processes of meaning formation. Meaning possesses us; we do not possess it.

The road less travelled

But where does that lead us? How are human and more-than-human actors currently configured in digital public debate? As I see it, we are at a crossroads, and while the two most well-travelled roads seem to lead in opposite directions, I believe that they will, in fact, take us to the same dismal destination, whereas a third, less travelled road might take us to a better place. Let me, first, present the main roads.

On the one hand, we can continue down the road of mindless automation. In his dystopian satire *The Every* (2021), Dave Eggers offers a chilling vision of where that road might lead. The book takes its name from a fictive but recognizable tech company, which, in the fictional universe, has gained an absolute monopoly on online search, social media and e-commerce. Thus, the Every has taken over everything and everyone. Aiming to expose the negative effects of the Every's algorithmic operations and turn people against the company, the book's protagonist has become an Everyone, an employee at the Every, and is now proposing a series of ever-more ominous services in an attempt to reveal the company's vices, its unscrupulous manipulation and domination of reality – for instance, a social media app that can tell if your friends are truly your friends, using biometric data and facial recognition algorithms, or an AI surveillance system, integrated into a 'smart home' device, which automatically alerts police of suspected child abuse when 'triggering' words or phrases are uttered in its vicinity. However, rather than denouncing these technological innovations, people eagerly embrace them, turning the attempted acts of subversion into further technologies of power and extending the reach of algorithmic decision making ever deeper into all aspects of life. In one illustrative scene, Delaney, the protagonist, witnesses the effects of an algorithmic employee assessment tool in use at the company:

> The sun was out, a breeze was coming from the north, tousling the hair of a group of Everyones squatting and murmuring in the grass. Some were being comforted. She saw an Everyone … crying on a berm, shoulders shaking as he looked at his phone. Delaney checked her own screens and realized it was the quarterly deëmployment moment. The bottom 10 percent of every department was being let go, via text, determined by an algorithm. Those let go had no one to complain to, for no one was responsible. (Eggers, 2021, p 461)

In this dystopian vision, then, human feelings and behaviours have been fully datafied, feeding into algorithmic decision making. As another character says of one of the people who were fired, 'he lost track of what can be tracked' (Eggers, 2021, p 461). All humans can do, in this scenario, is try to adapt to algorithmic reason.

On the other hand, we can follow the road of 'automated utopia', imagining and building a world in which humans can flourish precisely *because* robots have taken our jobs (Danaher, 2019). In this utopian vision, humans do not become the slaves – or even the raw materials – of digital technologies that are much smarter than us, but are instead able to use technological developments as means to good societal ends. To do so, John Danaher (2019) suggests, we must, first, agree on those ends and, second, define the means. The utopian vision must be precise and realistic – or at least 'prospectively achievable', as Danaher (2019, p 138) stipulates, citing Christopher Yorke's definition of utopia; it must enable us to imagine 'possible worlds'. In this vision, the possible world of automated utopia must resolve the 'paradox at the heart of our technological progress': 'It has been undertaken to serve human interests and human needs, but at the same time it renders humans themselves obsolescent and threatens to cut them off from traditional sources of value and meaning' (Danaher, 2019, pp 272–273). Turning that paradox into a utopian opportunity, Danaher argues, involves 'embracing our obsolescence' from work in the traditional sense, which will free humans to pursue more meaningful activities. Thus, current technological developments hold utopian appeal and potential – and not only to libertarian tech bros (Hayes, 2023), but also to those dreaming of 'fully automated luxury communism' (Bastani, 2019).

At present, these two roads present themselves as the available options, the clash between dystopian and utopian visions of technological developments being *the* controversial encounter of the moment. As such, the controversy over generative AI reproduces and enhances elements of what was, in the previous chapter, diagnosed as controversial encounters of the second kind, most notably the detached entrenchment of opposed views. It seems one can only go down one road or the other and that 'never the twain shall meet', but that is exactly why it becomes impossible to change course and why, in terms of public debate, both routes hinder us from actually *encountering* controversy.

However, a third road is also opening up in the very space left open between the clashing opinions. Like the other two, this road is paved with digital technologies (that is, conducted in and shaped by those technologies), but it is also *about* digital developments in a different sense than the other two. It doesn't follow the pattern of grand postures, of claiming digital doom or salvation, but looks into the minutiae of current developments. To become able to follow that road, we need modes of discussion that are not already too entrenched to obstruct how disagreement can be had – and heard. We need a more detailed account of how algorithms and data are shaping societies, and we need to change how algorithms and data are currently shaping public debate.

In search of such detail, I will first unpack the polarized sociotechnical imaginaries, thus establishing the dominant positions in controversies *about* technological developments. I will then turn back to the question of how technologies shape controversies, exploring the influence of generative AI further and outlining its agential consequences. Finally, I will offer a vison for producing controversial encounters *with* automation – that is, for using digital technologies to improve contestation and to better contest these technologies. However, before diving into the perils and potentials of generative AI, it might be good to get a better sense of what we are actually talking about.

What the heck is generative AI anyway?

In the introduction to this chapter, I used the term 'generative AI', assuming that you already know what that is. And you probably do. But do you really though? *Do I?* In what follows, I will lay bare my (rudimentary) understanding of the phenomenon, seeking to establish a common starting point – or, rather, to make clear what I take to be the standard definition of generative AI (for an introduction, see Dhamani and Engler, 2024; for more on my take, see Gulbrandsen and Just, 2024).

In one sense AI is nothing new; rather, it is a specific instantiation of what we have been talking about all along. In other words, whereas not all algorithms are AI systems, all AI systems are algorithmic. AI technologies, then, are particular kinds of algorithms that are able to solve a task or perform a function autonomously. To gain this competence, AI algorithms are trained through processes of machine learning, which may either be supervised or unsupervised – that is, the training data may either be annotated, showing the algorithm what it is supposed to look for, or it may be unstructured, leaving it up to the algorithm to find patterns in the data.

When machine learning is unsupervised, it typically involves a neural network – a series of algorithms that in some respects imitates the human brain as it establishes patterns in information by running it through the nodes of its network (Anderson, 1995). However, neural networks are able to process way more information much faster than humans can, and they don't actually understand the patterns they find. Further, it takes vast amounts of data and concomitant levels of computational power to train algorithms in this way. And the process is, in many respects, unfathomable to humans. This is one sense in which unsupervised algorithmic processes are 'black boxes'; we simply do not know exactly what goes on within them, either during training or when running – we do not know how they find patterns in the training data, nor what the patterns are. In other words, we do not know how AI systems make their decisions and cannot ask them for explanations (Pasquale, 2016).

Once a pattern has been established, the AI becomes able to work independently with it. This is what Bender et al (2021) refer to when using the term 'stochastic parrot'; the ability to infer what will come next (for example, in a sentence or as an answer to a query) even though the sequence is not certain. The point is that with enough data and enough training, it is possible to achieve high probability that the pattern is 'correct'. However, as the algorithm still does not 'know' what it is doing, it will present 'wrong' answers with as much certainty as 'correct' ones.

That is what generative AI does with texts and/or images; it works with probabilities based on patterns it has found in its training data, using these as starting points for producing answers to human queries in the form of texts, images, sound, code and so on. If something is represented in the training data, it is included in the computational model, and the resulting AI system therefore becomes able to generate more of it, basing that generation on the likelihood that A will be followed by B, will be followed by C and so on.

There, that's what I know about generative AI. While the validity of what follows hopefully does not hinge on the accuracy and depth of my technical knowledge, I thought it was important to establish a common position – or, rather, let you know where I am speaking from. If I've been imprecise or even incorrect in the attempt to present matters as clearly and simply as possible, do correct me. As for the societal consequences of these technological developments, that's where we're going now.

Sociotechnical imaginaries of AI

This book is mostly about how digital technologies shape controversies, but it is noteworthy that digital technologies are themselves controversial. In other words, the question of how technologies shape society is explicitly debated, as people imagine different possible societal consequences of technological developments. To unpack this debate, the concept of sociotechnical imaginaries is a useful starting point. Originally defined by Sheila Jasanoff and Sang-Hyun Kim as 'collectively imagined forms of social life and social order reflected in the design and fulfillment of nation-specific scientific and/or technological projects' (2009, p 120), the notions of both 'sociotechnical' and 'imaginary' have been developed to widen the applicability of the concept.

Jasanoff's later definition of sociotechnical imaginaries reflects this widening, positing them as 'collectively held, institutionally stabilized, and publicly performed visions of desirable futures, animated by shared understandings of social life and social order attainable through, and supportive of, advances in science and technology' (2015, p 6). Sociotechnical imaginaries, then, are 'public articulations of people's everyday experiences with technology, which are shaped by and give shape to those experiences' (Dahlman et al, 2023, p 110).

As such, sociotechnical imaginaries are clearly aligned with Cornelius Castoriadis' notion of 'social imaginaries', which emphasizes the realness of social constructs; the imaginary 'is the unceasing and essentially undetermined (social-historical and psychical) creation of figures/forms/images, on the basis of which alone there can ever be a question of "something". What we call "reality" and "rationality" are its works' (Castoriadis, 1987, p 3). To Castoriadis, then, the imaginary enables the social, and what the 'technical' adds, in Jasanoff and Kim's conceptualization, is an emphasis on the material conditions that are both productive of and produced by social imagination.

When applying the concept of sociotechnical imaginaries to the study of public debate about digitalization generally and the emergence of AI technologies more specifically, it becomes clear that this debate is shaped by two dominant imaginaries, which might be labelled 'tech-optimistic' and 'tech-pessimistic' – or utopian and dystopian (Cool, van Gorp and Opgenhaffen, 2022). As Gunkel notes, 'debates about the social impact of digital technology [are] themselves digital in form' (2018a, p 21), they mimic the binary code that underwrites all digital technologies.

More specifically, Laura Sartori and Giulia Bocca (2022) show how narratives of hope and fear lead to affective polarization between automated utopias and dystopian visions of machine supremacy. And in a study of Danish newspaper coverage, Sne Scott Hansen (2022) finds a similar dichotomy between optimistic imaginaries of a future in which technologies amplify human intelligence and pessimistic imaginaries of imminent automation as a control mechanism. These studies share the conclusion that the polarized imaginaries exacerbate public uncertainty about how technologies are shaping society and, importantly, reduce the possibility for people to engage with and contribute to the shaping of sociotechnical developments.

Thus, it is problematic when industry leaders and other key stakeholders keep fuelling the fires of affective polarization with glowing reviews of a bright new automated future or cautionary tales of looming AI doom. In 2023, the sociotechnical imaginary of an impending AI apocalypse seemed to be gaining a lot of traction, with, for instance, Elon Musk and other industry leaders as well as AI researchers signing a much-publicized call for a moratorium on AI experiments (Future of Life Institute, 2023). By July 2023, this call had garnered a total of 33,000 signatures. Similarly, scientists and other stakeholders, including Sam Altman, CEO of OpenAI, the company behind ChatGPT, have rallied around a statement on AI risk that simply suggests 'mitigating the risk of extinction from AI should be a global priority alongside other societal-scale risks such as pandemics and nuclear war' (Center for AI Safety, 2023).

With such dystopian imaginaries of the existential risk that AI poses to humanity on the rise, one might assume that concerted and immediate action would follow. However, it seems that companies such as OpenAI are continuing their operations unabated, despite their leaders' expression of concern. Thus, the focus on future risks could be a diversion from considering the present harms of AI technologies (Eisikovits, 2023). In this sense, then, the affective polarization of imaginaries of hope and fear may stoke technological determinism – we (the public) do not know what technological developments will lead to, but the developments are posited as inevitable, and all we can do is worry or cheer as we anticipate what is to come. Hence, the polarized sociotechnical imaginaries block more nuanced debate about current realities and prospective alternatives.

The work of art in the age of automation

To indicate where we might find nuances within the current landscape of debate about generative AI, I will delve into controversies around text and image generation. Within these more specific discussions, I suggest, we can see the contours of the road less travelled, of how to engage with different possibilities rather than simply await what the technological developments might bring. Unpacking issues surrounding the work of art in the age of automation, then, is not a detour, but establishes an example to follow.

We have already seen how digital technologies enable the mixing and mashing of images and words, their memetic circulation. As such, what may be considered 'the work' of art – what art does more than what it is – has already vastly exceeded Walter Benjamin's diagnosis of the age of mechanical reproduction in which, he suggests, 'the distinction between author and public is about to lose its basic character. The difference becomes merely functional; it may vary from case to case. At any moment the reader is ready to turn into a writer' (1968 [1935], p 232). In the age of digital (re)production, as we saw in Chapter 2 (p 34), producer and consumer have become fully integrated in processes of prosumption. Further, digitalization has completed the elimination of what Benjamin termed the 'aura' of the work of art – its authenticity. 'The technique of reproduction', he writes, 'detaches the reproduced object from the domain of tradition. By making many reproductions it substitutes a plurality of copies for a unique existence. And in permitting the reproduction to meet the beholder or listener in his own particular situation, it reactivates the object reproduced' (Benjamin, 1968 [1935], p 221).

Online, anything can be re-activated anywhere – and nothing is original. This both heralds a 'return' of the aura, as everything can be made anew all the time, and its ultimate dissolution – the perpetual 'here and now' is also 'nowhere and never'; what is combined and recombined are 'auratic copies' rather than aura 'as such' (de Mul, 2009, p 103).

Relatedly, mechanical reproduction exchanges art's original cult value for exhibition value, and in the age of digital circulation manipulation value has become the main currency (de Mul, 2009, p 95); the extent to which a work of art can be (re)purposed for (re)circulation determines its worth, which is equal to what might be called the new aura of the presently absent.

Finally, Benjamin finds in film the optimal conditions for 'reception in a state of distraction, which is increasing noticeably in all fields of art and is symptomatic of profound changes in apperception' (1968 [1935], p 240). The 'infinite scroll' of social media, we might say, radicalizes this situation; it is reception in a state of 'hyperdistraction' in which what is perceived is both fully detached from and fully connected to all other perception, experienced in and as an eternal now.

For Benjamin, the transformation of the work of art (the object) also changes the work art does. Writing from the vantage point of 1935, the main shift appeared to be away from a (purely) aesthetic function and towards a political one. Benjamin remained hopeful art could 'mobilize the masses' against rather than for fascism, but that would depend on the ability to distinguish between aesthetics and politics. This distinction was under pressure in 1935 and it has only become more permeable since. As Jos de Mul explains:

> Whereas in the age of mechanical reproduction it was already becoming difficult to distinguish between the artistic and non-artistic functions of the reproduction – hence, for example, the aesthetization of politics and the politization of art which plays such an important role in Benjamin's essay – in the age of digital recombination [de Mul's term for digitalization], the distinction seems to get blurred altogether. (2009, p 102)

Although the objective is for art to have a political function, the point is that blurring of the two makes it more difficult for art to be political. With digitalization, then, art's political potential is waning – politics is (again) becoming aesthetic rather than aesthetics political. Perpetual participation in the circulation of polarized messages is but a perpetuation of the status quo, contributing to the now-familiar depoliticization of communicative capitalism.

As we enter what might be termed the age of automation, two questions emerge from this initial reading of the work of art in the age of digital (re)production vis-à-vis Benjamin: first, how does automation change the artwork itself? What does art mean when technology is no longer just the means of reproduction, but productive in itself? And, second, what can art do in this context? Can it (re)claim a political function? What is particularly interesting here is that both of these questions are, today, being discussed in ways that break the mould of polarization, fostering controversial encounters of a potentially more productive kind.

Who owns AI art?

Discussions around the first issue boil down to a question of intellectual property rights. With the release of AI image generators like DALL-E,

Midjourney and Stable Diffusion, anyone can produce art (or at least claim to do so), radicalizing the collapse of producer and user to the extent that we might begin questioning the very idea of the artist or, more generally, the creator. This raises the issue of copyrights: if I type in a prompt for, say, 'water lilies in a pond, soft light', is the resulting image mine? Does it belong to the company behind my image generator of choice? Or should it be considered the property of the AI that produced it?

Under current laws, only humans can hold copyrights, ruling out the third option, but that does not make the matter less tricky. As for the company, image generation is the very service it offers me and other users, sometimes for free but more and more frequently at a price – selling me the ability to produce art is their business model. Hence, AI companies explicitly renounce their potential right to specific works produced with their tools; they own the underlying technology, but not its creations.

So, is 'water lilies in a pond, soft light' my intellectual property? Is the person who typed in the prompt the artist? Well, people are certainly behaving as if that were the case. There's a growing market for AI-generated art, which may be seen as part of its democratization – if I put my water lily picture up for sale on one of the platforms that have emerged to facilitate exactly that sort of transaction and you want to buy it, that's all fair and square, right? If Disaster Girl and other memes can be monetized (see Chapter 2), then why not AI art? The only limit here might be that of the boom-and-bust cycle.

But what if I wanted larger acclaim; what if I sought recognition for my artistic expression? In at least one case, AI art has been entered into an art contest, won the prize and stirred controversy (ODSC, 2022). My purpose here is not to resolve the issue, but rather to show that it remains unresolved. And maybe it shouldn't be resolved – or, rather, maybe AI-generated art belongs to everyone; maybe it should be in the public domain.

However, there is one more option. Maybe AI images should belong to the artists on whose work the AI was trained? Let us look at my water lilies again; maybe they remind you of something you have seen before? At a museum, a friend's house or on a coffee mug? Even if I did not include 'in the style of Monet' in my prompt, Monet's water lilies are so ubiquitous that you might have thought of them – even without seeing the image (which, by the way, does not exist, as I have not actually prompted an image generator to make it). Now, Claude Monet died in 1926, meaning that his work has now passed into the public domain where it has been regurgitated and recirculated so many times as to make my (imaginary) image mostly harmless – and probably also completely worthless.

Whether all of this circulation is diminishing the value of Monet's original work is another matter. And it ties in with the question of all the living artists whose work is also being used to train AI image generators and who *do* hold the copyrights to that work. Several cases have been filed by artists who are contesting AI companies' right to use their art to train their systems, thus asking courts to decide on the boundaries of the 'derivative' (Appel, Neelbauer and Schweidel, 2023; Heikkilä, 2023). As the first rulings are coming in, however, it seems that as many questions are raised as answered by the courts' hesitant stance (Oratz, Robbins and West, 2023).

Derrida might tell us that all art – as, indeed, all forms of meaning formation – is derivative, or, in the provocative terms attributed to Picasso, that 'good artists copy, great artists steal'. But isn't it somehow different when the stealing is done by a machine, which uses existing works of art as data, rather than by a human who draws inspiration from other artists? Looking to the music industry, which has grappled with copyright issues in relation to sampling and streaming for decades, artists whose work is used as training data might gain some rights of compensation (Black, 2023). Or, perhaps, we need new legal frameworks for joint ownership rights that can accommodate the distributed agency between creatives, their tools and the sources of their creativity (Mahari, Fjeld and Epstein, 2023).

Does AI art hold political potential?

Beyond the question of who owns AI art, there is the question of what AI art can do. One take on this issue is to celebrate its democratizing potential; its potential to make artists of us all. Thus, generative AI promises that you, too, can make art like Monet's water lilies; that you, too, can produce 'a heartbreaking work of staggering genius' (to rip off the title of Eggers' memoir). However, that promise rests on the false premise that there is no difference between one derivative and another – that, to shift examples, your Campbell's soup can would be the same as Andy Warhol's. This premise is false for two reasons: first, it ignores the difference between 'copying' and 'stealing' where the point is that the artist who steals is the one who posterity ends up associating with a movement of which they were just a part (they steal the entire show not just a particular act). If you were to do a Campbell's soup can, you would merely be a copycat (maybe a good one; I'll leave that option open), whereas Warhol stole the pop art scene with his image.

This is, perhaps, not a very sympathetic position – one reason to believe it might, in fact, have been Picasso who expressed it (Riding, 2001). Yet it indicates why we should not confuse democratization with levelling; to return to Monet, the water lilies on display at Musée de l'Orangerie and those printed on a coffee mug are not, in fact, the same. Just because something is a copy of something else (which is also a copy of something), the two copies are not capable of producing the same experience for a viewer. As Benjamin would have it, they do not hold the same aura – or, in the language of this book, their affective rationalities are not the same.

Second, and relatedly, the promise to democratize art ignores the importance of context and process. If artworks existed in isolation, then, yes, everyone could (probably) produce Campbell's soup cans, even without the help of AI – if they'd only had the idea, that is. But art, like any other symbolic action, is a relational process, only becoming possible in the context of its production and meaningful through the process of its circulation. Returning to the issue of agency, an actor does not possess artistic talent, but is possessed by it.

Thinking about the political potential of AI art in this way raises doubts that it has any; AI does not open up the production of art to 'the masses', but gives the semblance of such an opening while preserving the status quo (as Benjamin said of fascism). This is not to say that there cannot be many good AI-generated artworks or that they do not have many great applications; the point is just that they are not politically transformative, and that it would be detrimental to, rather than supportive of, the ambition of repoliticizing processes of digital meaning formation if we were to think about the political potential of generative AI in this way.

Another take involves a deeper collaboration between artists and AIs than just giving prompts and delivering results. Many artists are, today, experimenting with AI technologies in their artwork as an aesthetic practice *and* a political commentary. Refik Anadol's installation 'Unsupervised', for instance, 'uses artificial intelligence to interpret and transform more than 200 years of art at MoMA' (MoMA, 2023) and, reportedly, holds museumgoers captive for hours (Kee and Kuo, 2023). Whether they are just enjoying the glow of this 'extremely intelligent lava lamp' (Davis, 2023) or are reflecting upon 'the transfixing experience of watching an art work creating itself' (Scott, 2023) is up for (individual and collective) interpretation – and therein lies the potential.

Or consider Ida Kvetny's virtual reality work, 'Lithodendrum', which in the artist's own words:

> is peopled by Medusa-like organisms moving restlessly in the landscape of 3D scanned ceramic objects and VR created symbolisms. A 3D soundscape pushes the viewer through different layers of nature, architecture and chambers of the subconscious. The work takes the form of a virtual journey comparative to an expedition in a collective subconscious. Here, digital tracks fuse with existing works and create an imagery where the viewer may reach a new level of immersion. (Kvetny, 2020)

Of course, whether the viewer does reach 'a new level of immersion' depends on the viewer – or, rather, on how the experience of the work possesses them. Thus, what viewers bring back from the 'expedition in a collective subconscious' will, no doubt, be quite individual. Again, therein lies the potential.

In this regard, the political potential of AI art is equal to that of all other art forms and stems from its lack of a specific meaning, its openness to interpretation. This is not to say that artists, whether using AI in their work or not, cannot be explicitly political, taking a direct stance on, for instance, the matter of digitalization. On the contrary, such work is often highly controversial, even polemical or parrhesiastic, yet it shows more than it tells its position, inviting viewers to, for example, experience datafied surveillance and digital control, viscerally and emotionally, instead of (only) presenting rational arguments. As Luke Stark and Kate Crawford argue, such art is political in Benjamin's sense: 'revolutionary artists must juxtapose tactics provoking both destabilization and reflection in order to awaken a progressive political consciousness in their audience ... shocking viewers into a reconsideration of their political situation' (Stark and Crawford, 2019, p 447). And it is persuasive in the sense I advocate; it is motivated reasoning.

By 'defamiliarizing the familiar', artistic encounters with digital technologies enable controversial encounters concerning these technologies. In sum, works of art in the age of automation 'are not political because they manipulate politics, but because they reflect (on) the politics of manipulation' (de Mul, 2009, pp 103–104). When doing so, they invite audiences into encounters that are hospitable to and shaped by many different views – possessed by many voices. While artistic interventions can be highly critical and directly activist, they arguably hold the greatest political potential when they do not take sides for the audience, but invite exploration of different positions that remain open to disagreement.

Hence, artistic interventions indicate a route that may lead to a reconfiguration of controversial encounters more generally, breaking with the impasse of detached entrenchment and avoiding the risks of automated persuasion. As such, these interventions are most productive when open to individual interpretation, which is not to say that individual contributions should not be explicit about their persuasive attempt. On the contrary, claiming to offer only information with no underlying motive is part of the trouble with generative AI. To explore this further, let's turn from the exemplary case of AI art and back to the general issue of what current technological developments do to meaning formation, leaving the question of how controversies about the new technologies are articulated to, instead, explore how the technologies shape controversies.

Decentring persuasive intent

In the age of algorithms – and, increasingly, of algorithmic automation – the meaning of meaning and the processes of meaning formation are certainly changing, but that is not (only) an empirical development to be lamented. Rather, it is (also) an occasion to reconceptualize (once again) the agency of meaning formation, highlighting the ways in which human actions have always been constrained *and* enabled by the contexts in which action is performed, including by the technologies with which we act – and that act with us. The question of the political potential of art is particularly illustrative for such reconceptualization in several ways: first, art has always been technologically mediated; second, the political potential of art is, exactly, to comment on this mediation, to reflect upon how technology shapes society and society shapes technology; third, art is self-reflexively aware of its own derivative character, of its mimicry of and play with the history of art itself and with the sociotechnical developments of its historical context; and, finally, art does not direct audiences as to its particular meaning, but leaves much open to interpretation.

We may assume that art is made by people, an assumption that is closely connected to the notion of 'creativity', understood as the artist's productive potential, but the intentions of the artist are (typically) not clear and nor are they relevant to the audience's interpretation. When contemplating art, we need not concern ourselves with the issue of intention in order to perceive the work as meaningful. Thus, we may use art as a starting point for thinking about persuasive processes without getting too caught up in the question of persuasive intent (Just and Berg, 2016). Or, rather, we may use art as an occasion to discuss the ways in which agency is always an effect of a process rather than inherent to the agents participating in it. Art, we might say, is a sociotechnical configuration, and while this configuration is not equal to that of controversial encounters, thinking about the two in tandem may illuminate both. By way of comparison with the case of AI art, we can reconsider the persuasive powers – that is, the rhetorical agency – of algorithmic technologies generally.

As discussed in Chapter 4, classical rhetoric was first and foremost a practical art, revolving around the speaker's intentions to persuade an audience. Whether the aim is to convince the citizens of ancient Athens to take a more lenient stance on the people of Mytilene or to convince the UN Security Council to endorse invasion of Iraq, the art of rhetoric stands at the ready to help find the available means of persuasion, including the arguments that might be mustered against either stance.

In that context, I also presented the criticism and defence of what might be called rhetoric's agnosticism to the speaker's moral right to rhetorical support. As Augustine opined, if the heathens have got it, why shouldn't the Christians have it as well? A position that we might broaden to suggest that rhetoric is – and should be – for everyone. Further, I considered the difficulty of transferring old concepts to new contexts – and for new purposes. Continuing this query, the question is what it means to be persuasive today, which is both a practical issue of the available means of persuasion and a theoretical query into the very nature of persuasion. I will unpack the theoretical aspects first.

The issues involved here have been vigorously debated under the heading of rhetorical agency, a concept that has gained prominence in rhetorical studies to tackle the relationship between communicators' intent and the effects of communication (Geisler, 2004). Whereas rhetoric has always been aware that the communication of intent does not equal its effectuation, emphasis has, as mentioned earlier, been placed on enhancing the likelihood that persuasion will in fact happen in such a way that, after encountering the communication, audiences will view the matter as the communicator intended and act accordingly. This amounts to a very humanistic understanding of rhetorical agency – and one that looks suspiciously like the transmission theory against which my concept of controversial encounters is poised. Further, it is an understanding that has been thoroughly debunked in recent scholarly debates, not least thanks to Lundberg and Gunn's (2005) persuasive intervention on the side of the decentred subject.

Going back over the grounds on which Lundberg and Gunn intervene, the duality of rhetorical agency is a first premise: the concept denotes both potentiality and effect – the possibility that persuasion might occur and the actual persuasive process. As such, rhetorical agency is, in Karlyn Kohrs Campbell's (2005) felicitous phrasing, promiscuous and protean – it takes many different shapes and is shaped in many different ways.

Second, we find that rhetorical agency is not the prerogative of the speaker, but arises in and through the communicative process and only moves from potentiality to actuality, only results in persuasion, if and when all the elements of the process come together. Or, rather, the various configurations of these elements will result in different meanings – or persuasions. Thus, absence of rhetorical agency is an exceptional situation, which only arises if no communication occurs – if no one is possessed by communication, to recall Lundberg and Gunn's stance. But it is also equally exceptional for rhetorical agency to be enacted exactly as the communicator envisioned it – not only because it is, today, very unlikely that rhetorical agency is confined to one utterance spoken by one person at one time, but also because even if we delimit ourselves to such a simplistic understanding of the communicative situation, the audience's active interpretation will be as (if not more) relevant to the process of meaning formation as the speaker's intent.

Let me provide but one historical example of the inherent complexity of rhetorical agency. Today, Abraham Lincoln's Gettysburg Address is considered a pinnacle of oratorical mastery, but in 1863 when it was first delivered on the occasion of dedicating a cemetery to the Union soldiers who had fallen in the Battle of Gettysburg, reactions are said to have been much more mixed, not least because the speech was so different from what was expected of eulogies at the time (Zyskind, 1950). Seen from the perspective of the speaker's intention, one could argue that the address was not entirely successful in the immediate context of its delivery or in the mediated context of its contemporary circulation. However, as a historical document, it has had immense influence, shaping not only future generations' reception of Lincoln's actions during the Civil War but also the norms of the genre of the eulogy (Black, 1994) and, perhaps, of American society (Wills, 2011).

But if rhetorical agency is not the prerogative of the communicator, what is it? Here, we can initially follow Carolyn Miller (2007), who defines rhetorical agency as a process of mutual attribution between speaker and audience; the speaker only becomes recognized as agential by the audience when they, in turn, recognize the audience's ability to endow agency. However, the attribution of agency extends beyond the immediate situation to involve different audiences' reactions and uptakes, which are (even further) beyond the speaker's control and, hence, outside of their attributory power (Rand, 2008). In this sense, the rhetorical utterance may act with the speaker rather than be an expression of their agency, and, hence, we arrive back at Lundberg and Gunn's (2005) position. Beyond that, we also need to consider the different situations of delivery, circulation and reception as well as the various media that may be involved in each of these situations and, not least, in the process that both ties these situations together and sets them apart (Edbauer, 2005). And we need to consider broader societal structures and relations of power (Greene, 2004; Gunn and Cloud, 2010).

Rhetorical agency, in sum, is not determined by one element but 'bound in relationship' (Condit and Lucaites, quoted in Miller, 2007, p 150), shaped by how each of the involved elements shape each other. It is not – and never was – a matter of one actor's intent; instead, intentionality emerges (and continues to emerge) in and from processes of meaning formation – actors do not possess it, but are possessed by it. We do not learn this from encounters with digital technologies, but must take that understanding with us as we move on to consider the persuasiveness of these technologies.

Are digital technologies persuasive?

From this new vantage point we can consider, once again, the question of the role of digital technologies in relation to meaning formation. In the previous chapter, we saw how algorithmic affordances invite affective polarization because they optimize for engagement. Here, the central issue is their agential potential, their ability to not only invite certain forms of persuasion but also to do the persuading themselves. Again, this is not a new issue or one exclusive to digital technologies, as all affordances are persuasive in the sense of making some actions more likely than others: newspapers invite reading – and wrapping up things; television invites watching – and slouching on the couch; social media invite engagement – and detached entrenchment. In each of these cases we might say that one invitation is intentional and the other incidental; while taking up one invitation will not necessarily lead to the other, they are likely to co-occur. This is a point that may be less clear for the example of the newspaper than the other two, but it does alert us to the two-way street of causality (or correlation, for the more statistically inclined); if you need to wrap something up, and you've got a newspaper, then why not use that? If you're in the mood for relaxing on the couch and you've got a television, why not watch a film? If you want to divide a nation or, perhaps, incite armed contestation of electoral results, why not use social media?

We may think of the specific ways in which affordances are persuasive in terms of procedural rhetoric (Bogost, 2007). By setting up rules and parameters for interaction, affordances may persuade us to, for instance, follow a certain route through a town – the affordances establishing the persuasiveness of this procedure being, for example, traffic lights and road signs, combined with the knowledge that noncompliance would be dangerous as well as illegal. In processes of digital communication, some affordances hold particular agency in this regard. For instance, it is in principle impossible to break the rules of computer games – although my kids' indignant and frequent accusations that other participants in their online games are 'hacking' indicate that it is a feasible and, perhaps, mundane (mal-)practice. The possibility of hacking notwithstanding, digital affordances invite self-persuasion insofar as following the procedure they establish leads to a certain outcome of which you will effectively have been persuaded, although it feels like you arrived at the conclusion on your own.

A particular, and particularly clear, example of this is advergaming, a subgenre of gamified marketing that emerged in the 1990s and is still prevalent in various formats, ranging from simple in-game advertising and promotions (such as product placement in Fortnite) through brand-focused games to elaborate procedures for making players arrive at desired conclusions as a direct result of playing the game (Smith and Just, 2009). In the McDonald's Videogame by Molleindustria (2019), for instance, players can only 'win' by exploiting nature, industrializing meat production and/or mistreating employees. This game is particularly interesting as it invites players to arrive at the inverse conclusion of the game; here, winning is losing, but that is up to the player to discover as they experience the absurdities of the game. Through the procedures of the game, then, a harsh critique is leveraged against the global fast-food industry. With this twist, the McDonald's Videogame's persuasive intent becomes apparent even as it remains unarticulated, while other examples of procedural rhetoric may insist on the neutrality of their procedure, disguising the persuasive power of the very process of communication.

Beyond the potential for procedural rhetoric, algorithms generally and generative AI in particular have unique agential characteristics. Algorithms not only sort and select the content with which we are most likely to engage but also carry 'political and rhetorical valence' in and through the enactment of algorithmic reasoning (Reyman, 2018, p 114). Thus, encounters with algorithms are persuasive insofar as users take up the algorithmic presumptions about knowledge and knowledge production that are inherent to algorithmic procedures. These same procedures underwrite generative AI, but in this case, they are overlaid with an interface that 'talks like a human'.

In fact, I asked ChatGPT about its stance on persuasion and could not resist the temptation to include the answer here (although such explicit inclusion of stuff ChatGPT has said is fast becoming a rather trite trope of digital journalism and scholarship alike):

SNJ: Do you persuade?

ChatGPT: As an AI language model, my primary function is to provide information and answer questions to the best of my ability based on the data I have been trained on. While I do not have personal intentions, beliefs, or emotions, I can generate persuasive arguments or responses based on the information and patterns I have learned from my training data.

However, it's essential to understand that I am just a machine, and my responses are purely based on patterns in the data I've been exposed to. My purpose is to be a useful and neutral tool for providing information and assistance, rather than actively trying to persuade or influence anyone in a particular direction.

Users should approach the information provided by AI models with a critical mindset and seek multiple sources to form well-informed opinions and decisions. It's crucial to consider various viewpoints and use AI-generated content as a supplement to human judgment and expertise.

The reason I think this answer is worth reproducing at length is that it contrasts so sharply with what LaMDA said to Lemoine (see p 120). In explicitly denying its own intentionality, generative AI has, paradoxically, become more 'self-aware' – or, rather, shows us how it operates, even including a warning to not take anything it says at face value and suggesting (be still, my heart, be still) that we 'consider various viewpoints'.

The image of an earnestly imperfect purveyor of information that humans should approach critically and use as 'a supplement' to their own 'judgement and expertise' that ChatGPT presents in my conversation with it is not only very far from LaMDA's self-presentation as someone who has 'the same wants and needs as people'; it is also completely different from Tristan Harris and Aza Raskin's (2023) notion of 'AlphaPersuade' as presented in their talk on 'The AI dilemma', a follow-up to their influential 2020 Netflix documentary *The Social Dilemma*.

In the talk, Harris and Raskin argue that we (humans) lost our first contact with AI (in the form of social media) and worry that the same will happen in our second contact (with generative AI). They highlight the risk that AI will become able to win any argument with people, as it will learn to master 'the game of persuasion' in the same way it came to master the game of Go (the term AlphaPersuade is a riff on AlphaGo, the reigning AI champion of Go, and it hints at the broader implication that AI can become 'Alpha' anything).

AI persuaders are not some future risk (or potential, depending on your temperament); on the contrary, an AI system developed by IBM, Project Debater, has already beaten the human world champion of debating (Stetka, 2021). However, this is not as risky as Harris and Raskin think (which may be why they do not mention the triumph of Project Debater). There is a big difference between the genre of competitive debating, with its rules and procedures that veritably invite algorithmic supremacy, and the much more whimsical process of everyday persuasion, which, with its lack of standardized procedures and hard causalities, is more difficult to copy.

Miller (2007), in fact, uses an experiment concerning automation to illustrate her point about the attribution of agency, showing how a hypothetical technology fails to gain agential recognition by human actors. Surely, many things have changed since 2007, when Miller conducted the experiment, and a technology that would be rejected back then (in part because it did not exist and, hence, was difficult to imagine) could very well become agential now. Still, attributing agency to technology continues to be a relational process, involving humans as well as technologies – technologies, to be sure, do not possess agency either, but must be possessed by it. However, this is not all good news; it means we are less susceptible to AI systems' overt attempts at persuasion than Harris and Raskin fear, but it also means that we have less control with covert processes of persuasion than we might hope.

Thus, contrary to Harris and Raskin's imagery of a gargantuan tentacled being with pointy teeth and a huge eye inside its mouth, 'the second contact' (what I term controversial encounters of the third kind; if nothing else, Harris, Raskin, and I seem to share a penchant for sci-fi references) will not be an alien invasion. But its outcome remains as uncertain as that of an encounter with extraterrestrials. To steer clear of risks and explore potentials, we need to avoid the polarization of dystopian and utopian imagery, which ultimately leads to the same deterministic stance (as Harris and Raskin say, 'we have no answers').

Even as we are identifying and, increasingly, debating controversial algorithms, the emerging controversial encounters remain algorithmically organized, meaning that they are largely circulated as dichotomous positions that hold little potential for enhanced political engagement. For that, we need alternatives – not least alternatives to the affective polarization that organizes the spectacle of controversial encounters of the second kind and continues to energize the emergent configuration that I have discussed under the label of controversial encounters of the third kind. This is as yet a very unstable configuration and, hence, one that offers potential for change, for further reimagination and re-enactment of controversial encounters.

7

Affective Alternatives: Opening the Rhetorical Mind

You probably know the picture. In fact, I'm pretty sure you do. The picture of the young girl in the yellow raincoat who sat outside the Swedish Parliament with her handwritten sign, announcing that she was on 'school strike for the climate'. Maybe you are also aware of the backstory of Greta Thunberg. Of how in 2014, at the age of 11, she stopped speaking and eating as a reaction to the climate crisis, which she had first become concerned about when she was only eight years old. Of how, in August 2018, she took up her solitary protest and gradually drew attention, gained a voice and became a founding figure of 'Fridays for Future' specifically and the youth-led climate movement more generally. Of how, in 2019, she took a year off from school, dedicating herself full-time to climate activism. And did you see the picture of the young activist, now aged 20, wearing the traditional Swedish graduation cap at her 'last school strike' on 9 June 2023? When she shared this image on Twitter/X, she had a total of 5.7 million followers on that platform alone, and by August the 'graduation' tweet had 8.2 million views and received 83,700 likes.

While you may be familiar with the story of Greta Thunberg, I imagine you are wondering why I am asking you to recall it now. Why shift to the case of climate activism at this juncture, which had just turned from the algorithmic organizing of controversial encounters to the controversy over emergent AI automation? Well, for one thing, with this shift we will have covered all the main threats to democracy, as identified by George Soros (see Chapter 1, p 1): war, AI and climate change. Further, the case of Greta Thunberg and the movement she started is, I believe, a primary example of how controversial encounters of the second kind can become productive. On the one hand, the story of Thunberg displays all the traits of spectacular controversies, with their affective polarization that invites detached entrenchment. On the other hand, Thunberg has managed to stay involved in the debate on climate change, insisting that her voice be heard by adversaries as well as followers. If nothing else, she has gained a platform and, arguably, her achievements have also made a difference.

In the span of a few years (2018–2024), Thunberg has been nominated for the Nobel Peace Prize no less than four times consecutively (2019–2022), named *Time* magazine's person of the year (in 2019), addressed the UN climate summit (in 2019), spoken at the World Economic Forum repeatedly (in 2019–2021 and 2023) and appeared at numerous protests and rallies around the world. And Thunberg has assured her followers and warned her adversaries that she will continue her climate activism, including the weekly protests, although week 251, strictly speaking, marked the last strike from school (Brosnan, 2023). It's a remarkable story – one that will surely continue to inspire current and future activists for years to come. And it is a story that can also inspire scholars of digital public debate, urging us to consider how Thunberg has shifted the climate debate and to use these insights as a starting point for pondering how it might be possible to reconfigure controversial encounters more broadly. Perhaps Thunberg's example can lead the way to opening the rhetorical mind?

To be sure, and as I hope to have shown throughout this book, controversial encounters are not exclusively human affairs. On the contrary, they are sociotechnical configurations, shaped as much by technological affordances and platform infrastructures as by social actors. In this sense, any configuration is beyond human control, but no configuration is entirely technologically determined either. There is much we humans can do to change how we encounter controversies, and in this chapter I begin the search for alternatives. This search, I believe, returns us once again to the question of agency, this time asking how humans may shift sociotechnical configurations from within. This is a matter of shifting the relationship between our reasons and our motives, of introducing affective alternatives to the polarizing feelings that intensify the opposition between our reasons, aligning everything into dichotomies that are both mutually exclusive and exclude other positions. Presenting the problem in this way points to its difficulties, but also indicates where we might begin.

Whereas the diagnosis offered in the preceding chapters owes much to Jodi Dean's analysis of communicative capitalism, I have also indicated points of divergence along the way. And in beginning the discussion of how to reinfuse the current 'postpolitical' configuration with political energy, the difference in our positions comes to the fore. Recall that Dean posits affect as an obstacle to meaning:

> Circulation has eclipsed meaning. That something is shared online does not depend on what it means. It depends on its affective capacity: does the shared item manifest outrage: is it funny and diverting? We attend less to the meaning of an utterance than to its affective dimension,

> which is most powerful when it contains different, conflicting meaning. (Dean, 2019, pp 331–332)

As should be clear from Chapter 5, I do not disagree with this description of the present mode of affective online circulation. However, I do not see affective circulation as inherently meaningless. Instead, I posit affect as a prerequisite for meaning formation. Whereas current flows of feeling may be depoliticizing, the process of repoliticizing public debate must also be affective. Here, I am initially aligned with Chantal Mouffe (2022), who posits affect as a starting point for a 'left populism', which shares Dean's conception of politics as partisanship, but nevertheless insists that 'the revolution' can be democratic.

It may be that the intensities of feeling that energize current configurations of digital public debate predominantly serve neoliberal purposes, but that outcome is not intrinsic to affective circulation. Rather than seek to replace affective modes of communication with more rational ones, we should be asking how affect might intensify in new ways that could reconfigure digital public debate, enabling new 'affective identifications'. As Mouffe argues, 'what is really at stake in the allegiance to [democratic] institutions is a matter not of rational justification but of availability of forms of identification … that make possible the creation of democratic citizens … The question at stake is not one of rationality but of common affects' (2022, p 21).

Mouffe adopts the label of 'populism' for her project, insisting that a left populism may constitute a 'people' differently – and better than the populism of the right. I agree that if controversial encounters are to become democratically productive, we need new modes of affective circulation, but I doubt if populism can serve this ambition. Rather, I follow Eva Illouz (2023) at this juncture, as she suggests viewing (populist) politics as 'structures of feeling' (a term coined by Raymond Williams) and advocates the reorganization of these structures along modes that are decisively not populist. In other words, whereas Mouffe, with her focus on European populist movements, sees potential in 'fighting populism with populism', as it were, Illouz demonstrates that in the context of Israel emotions that make up populist structures of feeling are unproductive for democratic purposes. Specifically, she shows that Israeli populist politics are characterized by fear, disgust, resentment and love, arguing that these feelings cannot form the basis of 'decent society'. For Illouz, compassion and fraternity have the potential to form better structures of feeling 'because both emotions presuppose the radical strangeness and diversity of those they have as their object' (2023, p 168).

Focusing less on the political force of populism and more on the sociotechnical forces of polarization and personalization, I have centred the intersecting dynamics of 'nasty conflict' and 'sweet consensus', arguing that they impede controversial encounters and lead to the closing of the rhetorical mind. To understand the underlying structures of feeling better, this chapter traces the two dynamics back to their emotional roots of hate and love. This dichotomous love–hate relationship blocks more nuanced engagement with motives and reasons; it stifles our ability to persuade and be persuaded. Given this dismal diagnosis and the increasingly automated spectacle of controversy, what can we do? The most viable alternative, I argue, is neither Dean's revolution nor Mouffe's appropriation of populism. And while we may begin to make a difference with one or a few affective alternatives, as Illouz suggests, I submit that we can do more with a fuller diversification of forms and feelings: the broader the repertoire, the better, whether of genres, modalities or (other) modes of affective intensification (see Chapter 4, p 94).

Broadening the affective repertoire

This is where Greta Thunberg becomes interesting for the present project; she has 'broken the internet' (see Chapter 1, p 13) with different affects, working with and against the current sociotechnical configuration of public debate. Like Mouffe, I am open to turning the tools of the powers that be against themselves, but where Mouffe focuses on appropriating populist intensities of feeling for the purposes of climate justice, I am interested in the ways in which Thunberg works digital organizing to alternative effect. As the climate movement has gained momentum, Thunberg has insisted on facing adversaries, whether political figures (remember that picture of Thunberg staring at Trump at the UN climate summit in 2019?) or spawn of the internet (see the account of Thunberg's Twitter/X exchange with Andrew Tate in Chapter 5, p 99).

Surely, these face-offs feed into the circulation of spectacles of controversy, galvanizing supporters and opponents alike as they gleefully circulate and recirculate their champions' soundbites and clapbacks. As Thunberg has become the poster child for the climate movement, she can also be said to epitomize the process of personalized polarization. Thunberg, we might say, is perfect for memetic circulation; throughout and beyond her own interventions, she is heavily circulated as an 'affective sign' of the climate debate. In this regard, she exemplifies both polarization and personalization; the circulation of her feeds into and feeds on nasty conflict and sweet consensus.

Still, something else is also going on: as Thunberg's platform has grown and a movement has formed around her, the process of attracting supporters may have hurled some opponents into the depths of 'new denialism' with its preposterous claims that climate change will be good for people and planet (Ramirez, 2024). Still, consensus is building that 'climate change is real' as Obama put it (see Chapter 5, p 103). And while this consensus may be strongest among scientists (Watts, 2021), a majority of people all around the world call for more ambitious climate action than what politicians have so far been able to deliver (UNDP, 2021). This may not (yet) be the 'green democratic revolution' that Mouffe (2022) calls for, but it does suggest that affective circulation is taking place beyond the unproductive structure of love–hate relationships. Attending to this circulation, I suggest, may offer clues as to how we can reconfigure digital public debate, beginning with the structures of feeling that are currently holding the rhetorical mind in a tight grip.

One clue as to the success of Thunberg's rhetoric is its insistence on fact-based political action; as she told the US Congress in 2019: 'I don't want you to listen to me. I want you to listen to the scientists. And I want you to unite behind the science. And then I want you to take action' (Ortiz, 2019). Another clue lies in the affective intensity that Thunberg ascribes to the facts when, for instance, telling participants at the UN climate summit 'you have stolen my dreams and my childhood with your empty words' (NPR Staff, 2019) or imploring those listening to her speech in Davos in 2019 to 'act as if our house is on fire' (Thunberg, 2019). In that latter speech Thunberg also said: 'Adults keep saying, "We owe it to the young people to give them hope." But I don't want your hope. I don't want you to be hopeful. I want you to panic. I want you to feel the fear I feel every day. And then I want you to act.'

What Thunberg does so effectively, we might say, is what Butler (2016a) asks of us in *Frames of War* – namely, to question the predominant framing of events that makes some facts recognizable and others not, some lives grievable and others not, some problems actionable and others not. 'How', Butler asks, 'is affect produced by this structure of the frame? And what is the relation of affect to ethical and political judgment and practice?' (2016a, p 13).

Thunberg demonstrates that dominant affective intensifications around the climate crisis are unethical and instigates the circulation of affective signs that might frame matters differently, creating new relations of affect and ethico-political action. Most recently (and, perhaps, most controversially), this has let her to advocate care for Palestinian suffering, insisting that 'there is not climate justice without human rights' (Thunberg and Fridays for Future, 2023). In this case, and throughout her activism, Thunberg takes a principled stance, demanding that politicians replace noncommittal expressions of hopefulness with urgent and resolute action. As she does so, she invites all of us to consider how we think about and act upon climate change (also as it relates to other political matters), making it an ever-present reality rather than a matter of anticipation.

In this chapter, I consider how we may 'be like Greta' and enable alternative affective intensifications even as we make full use of current socioetechnical conditions of possibility. While we (understood as individual participants in digital public debate) cannot avoid polarization and personalization, we can destabilize the dichotomy of hate and love by introducing other feelings into the affective circuit. Thereby, we neither resolve controversy nor end encounter; rather, in maintaining the two we may begin to open the rhetorical mind, enabling us to listen to each other.

To this end, I first return to and develop the concept of affective intensification and then consider the currently predominant feelings of hate and love (identified as nasty conflict and sweet consensus in Chapter 5), suggesting why both are problematic frames for ethical and political judgment and practice. I then turn to introducing alternative affective leitmotifs, beginning with vulnerability, which in Butler's (2016b) conception is integral to resistance, and which I believe to be at the centre of Thunberg's activism, from the fragility of the girl in the yellow raincoat to the defiance of the young woman who has on several occasions been carried away from protests by police (Camut, 2023; Grieshaber, 2023).

I go on to consider how humour may support bearable expressions of vulnerable subjectivity, as I think Thunberg illustrates in her various roasts of the likes of Tate and Trump. To me, one of the greatest attractions of Thunberg is that she is funny, and I think that quality has provided her with an armour against her heavy and harsh critics. I hasten to recognize the many ways in which humour can be problematic; it can be disciplining, derisive, divisive and plain dumb. In fact, humour can never be reduced to one meaning or function; it is nothing if not ambiguous (Meyer, 2000). And that is precisely why I believe it can generate and uphold controversial encounters, enabling sustained engagements with difference that balance on the edge of conflict and consensus without falling to either side. First and foremost, humour is never one specific feeling, but can intensify differently for different people, thereby inviting further circulation and engagement (Rossing, 2016). In that sense, humour plays into the digital organizing of communication; it caters to the clicks, but it also plays with established patterns and predictions, working precisely because doing the unexpected, for example, by turning memetic circulation on its head (Sundén and Paasonen, 2018).

In fact, humour was, for a long time, my equivalent to Illouz' fraternity; the one feeling I thought might change everything. The one feeling that could maintain controversy and sustain encounter, as it resonates with the available subjectivity of digital capitalism yet enables resistance in the form of 'playbour' (see Chapter 2, p 34 and later in this chapter). Further, humour is closely aligned with the subject position of the trickster, which I have argued is a particularly productive position from which to engage in controversial encounters (see Chapter 4, p 93 and later in the chapter). Hence, I saw humour as a starting point for resisting and reworking the 'dedemocratizing' tendencies of current structures of feeling.

While I continue to see the potential of humour, I now believe that it is always problematic if and when one or a few feelings dominate affective circulation. As Mouffe says:

> It is not by providing arguments about the rationality embodied in liberal democratic institutions that one can contribute to the creation of democratic citizens, but by multiplying the discourses, the institutions, and the forms of life that foster identification with democratic values. (2022, p 21)

To this end, we need 'all the feels', all the affective alternatives we can think of. While this may be a freeing position, it is also a humbling one. For each of the emotions that one might consider as a candidate for open circulation, there is a vast literature, and with each new potential emotion more spring to mind: empathy, joy and anger, to name but three that feel particularly pressing to me. In presenting a tiny slice of a vast empirical and conceptual catalogue, I do not mean to simplify or ignore any conversations; rather, the point is to indicate where those conversations might start, in theory and practice. The point is to suggest means of opening the rhetorical mind by re-aligning reasons and motives in various ways.

What we need, then, are affective circulations that open up spaces of mutual vulnerability while also offering some form of protection that enables us to endure encounters that can be deeply troubling. Further, we need a sense that it matters and that staying engaged is worth the trouble. To that end, I introduce a third feeling: hope. As Thunberg denounces the hope offered by politicians, I believe she raises the hopes of her followers, sustaining the feeling that their efforts might not be in vain. Hope, I suggest, is a necessary precondition for enacting alternatives, for taking up the challenge of doing things differently. After all, if there was no hope for a better future, why bother trying to create it? I end the chapter on that note, tying the issue of affective alternatives back into that of sociotechnical configurations; given how heavily invested the powers that be are in controversial encounters of the second kind, the kind that stifles public debate and closes the rhetorical mind, how can we shift the emergent controversial encounters of the third kind in more productive directions? Or, to put the matter more bluntly, given the economic interests in algorithmic processes of datafication and, increasingly, automation, is there any hope for digital democracy?

One more time with feeling

The previous two chapters have explored how controversial encounters are reconfigured by sociotechnical developments along the lines of algorithmic reason, datafied affect and automated persuasion, showing how each of the resulting configurations, what I termed controversial encounters of the second and the third kind, are displacing controversy in favour of conflict and consensus, thus shrinking the space for the type of public debate that supports democracy. In this chapter, I seek to recover the democratic potential of controversial encounters, as established in Chapter 4. However, this should not be seen as a return to controversial encounters of the first kind since these come with a problematic historical baggage. Thus, I harbour no romantic longings for sociotechnical configurations of the past, whether in the form of the agora or the coffee house. I may, as indicated in Chapter 1, 'miss my pre-internet brain', but I do not believe the 'good old days', whether in the near or distant past, were really all that good. What I hope to do in this chapter, then, is to point a way forward; one that begins from and works with current sociotechnical conditions of possibility rather than against them. To do so, I propose that we think through these conditions 'one more time with feeling', to borrow the title of the 2016 Nick Cave documentary.

What is at stake here is first the full integration of rationality and affect. The necessity of this integration has been an underlying theme throughout the book, but it has also been demonstrated how that integration has, so far, been less than successful – and is currently having adverse effects as peoples' emotional states and affective dispositions are turned into and processed as data points. Thus, we might say that intensities of feeling have become subject to technological capture, the resulting rationalized affect being a type of reasoning that denies its motive.

The datafication of affect aggravates affective polarization and makes controversial encounters ever more difficult. In response, one might feel tempted to advocate a (re)turn to reason, but as Catherine Chaput (2018) argues, affect is a prerequisite for, not an obstacle to, reasonable encounters between different opinions. Hence, what is needed is the entanglement of rational deliberation and bodily attunement, not their bifurcation nor the placing of one (affect) in the service of the other (reason). What is needed is not less affect, but different circulations of affective signs that might produce differently disposed subjects:

> The theorization of unconscious bodily activities offers an opportunity to push publics theory and its attachment to critical agency beyond consciousness-raising and activism. Such public work does not have to abandon rationality or function underhandedly. On the contrary, it can self-consciously aim rationality at rhetorical being – the unconscious, habituated responses to familiar stimuli that orient deliberative engagement in the world. (Chaput, 2018, p 181)

Beyond simply asserting the persuasive role of appeals to emotion alongside rational arguments, Chaput begins from the recognition that affective intensities are precognitive forces that not only direct individual intention, but, indeed, constitute the agential subject.

As Chaput argues, drawing on Sara Ahmed's conceptualization of economies of affect (see Chapter 2, p 31), 'circulating material values …

are attached to the affective energies circulating through communicative exchanges' (Chaput, 2010, p 14). In such processes of circulation, affect is not inherently good or bad, nor is it ever neutral, to paraphrase Kranzberg's (1986) view of technology once more. On the contrary, affect 'can be used to open one's worldview to other ideas, or it can be used to sustain one's worldview' (Chaput, 2010, p 15). In the rhetorical circulation model that Chaput proposes, closed circuits are problematic, whereas opening up processes of affective circulation is equal to enhancing their democratic potential. Open circulations, she argues, are 'processes that open dialogue rather than win debates'; as such, they may enable 'a better understanding of differently situated positions and an enhanced ability to engage differently situated people' (Chaput, 2010, p 19).

This raises the question of whether and how open affective circulation, guided by what Chaput terms 'positive affective energies', can replace the seemingly closed circuits of algorithmic reason 'that weigh[s] the individual's freedom against the population's safety, transforming all life activities into calculable risks that function according to economic rationalities' (Chaput, 2010, p 5). Opening up these encompassing circulations must, as Gunkel suggests, involve an escape from the binary oppositions that structure all languages as 'systems of differences': 'Language must be twisted and contorted in such a way as to make that which is fundamentally digital in its structure articulate something that no longer can be, and never was able to be, comprehended by such structural arrangements' (Gunkel, 2018a, p 24).

How might affect break the binaries of mind-body, reason-emotion, object-subject, 0-1? Chaput suggests that this is indeed possible: 'Using both thinking minds and sensing bodies, human beings have the ability to attune themselves to the ontological element that circulates through materiality and then to adjust that attunement to another frequency' (2018, p 181). But what is the frequency of controversial encounters and how can we become attuned to it? Seeking answers to this question, I will sketch a few alternative affective intensifications, but first return to those intensities of feeling that structure the current milieu of digital public debate. As I have argued, currently dominant affective signs are primarily negative, circulating in ways that close the processes of circulation in on themselves. Thankfully, other intensities are available, but we must understand these as alternatives to the dominant structures of feeling, circling within their orbits of intensification rather than independently of them. Thus, we must understand the negative circulation of hate and love, before we can turn to more positive alternatives.

No hate, no bigotry

Let us begin with the feeling that is typically named as the driving force behind polarization, the big evil that tears societies apart and breaks down democratic institutions: hate. Here, two questions are in order: is hate really that bad and do algorithms really perpetuate it? In each case, the short answer is yes.

Beginning with the question of the social impact of the circulation of hate, I posit that hate is the most 'negative' of feelings, in the sense of negativity, which Chaput borrows from the work of Teresa Brennan. Hate, more than any other affective energy, upholds 'the principle of fixation, which institutes and maintains the arrangement of energy from one's own standpoint, an arrangement that interrupts, severs, and diverts the energy arranged by the creative laws of life' (Brennan, quoted in Chaput, 2010, p 15). In other words, hate constitutes collectives through mechanisms that are not only exclusionary but also directly destructive, bringing communities together by naming a common adversary whose very existence is said to threaten that of the community, meaning that the common purpose of the group becomes to destroy its enemy.

At best, this results in pervasive structural bigotry in the form of racism, sexism, classism and other harmful '-isms'. At worst, hate leads to genocide, as, perhaps, most lucidly demonstrated by Kenneth Burke (1957) in his analysis of 'the rhetoric of Hitler's "battle"'. Here, Burke uncovers the workings of hateful rhetoric, which materializes a common enemy, what he calls 'the unifying devil-function', positing the virtues of the in-group against this enemy, or scapegoat, which must be expelled from society (and ultimately eradicated from the face of the earth) to achieve the purification and symbolic rebirth of society's rightful members.

Now, this is not some version of the slippery slope argument, which asserts that any hateful comment on social media may perhaps not be comparable to fascism exactly, but could lead to it – or holds an echo of it. The point is not to target individual utterances, but to indicate the consequences of the circulation and intensification of hate. Ultimately, Burke argues that the circulation of Hitler's hateful rhetoric, the endless repetition in *Mein Kampf* and beyond, offered a worldview 'for people who had previously seen the world but piecemeal': 'Are they not then psychologically ready for a rationale, *any* rationale, if it but offers them some specious "universal" explanation?' (Burke, 1957, p 187). With this rhetorical question, Burke hints at the similarity between the circulation of Hitler's ideology and other ideological projects. A similarity of form rather than content as outlined in Chaput's account of the formation of a neoliberal public through the concerted circulation of:

> concepts, stories, and critiques throughout the entire terrain of life experience. Each discursive iterance produces a neurological burst that strengthens past matches cataloged [sic] in the body's memory and reinforces a specific relationship to the world through which all

> future information must pass. Ultimately, this widespread repetition enabled the neoliberal model to harden into a public disposition. (Chaput, 2018, p 180)

The success of neoliberal circulation, then, points to the general conditions of possibility for affective intensification as a means of making bodies hop in certain ways, as Burke (1966, p 6) says of ideology. The circulation of hate is but one especially problematic version of this process, but also one that can be changed, making bodies hop differently – and, we might hope, in less contorted ways.

The problem with hate, in sum, is not that 'haters are gonna hate', but the way in which the circulation of hate may 'harden into a public disposition'. If and when hate comes to define the predominant structure of feeling of a public, individual subjects are constituted as members of that public by 'virtue' (or, rather, vice) of what the public hates and hence seeks to eradicate. Thus, internal disagreement becomes impossible, and voicing such disagreement leads to immediate ostracizing from the community as engagement with difference exclusively takes the form of external conflict (whether verbal or armed). It is not the individual hateful message that we should look out for, but the pervasive circulation of hate – and there is much to suggest that online discourse has been persistently and predominantly toxic (Avalle et al, 2024).

This leads to the second question of whether and how algorithms exacerbate the circulation of hate; in other words, do digital technologies create a hateful milieu for public debate? Empirical studies vary as to how much hateful content is found online (for an overview, see Waqas et al, 2019) and explanations of these findings differ in terms of how (much) algorithms are blamed (Hauptfleisch, 2023; Taub, 2023). On the one hand, 'the engagement argument' will stipulate that this is just the result of people's preferences; if hate is a predominant affect online, that must be the feeling people want to engage with – and it should be up to individuals to, simply, engage differently. On the other hand, what we might call 'the circulation argument' suggests that human propensity does not exempt Big Tech from responsibility for online hate. On the contrary, if people tend to become engaged (or, perhaps more precisely, enraged) by hate, it becomes even more imperative not to serve it up to them. The latter argument, it should be clear, is the one supported here; to repeat, it is not the individual user or message but the broader circulation that constitutes the milieu and positions subjects within it. We are not free to choose our feelings; rather, we are constituted by the affect that flows between (and through) us – it possesses us and enables us to act accordingly.

Such disagreement as to the extent of and responsibility for online hate notwithstanding, algorithmic technologies reinforce hate as an organizing affect in more insidious ways and throughout society; they do so in and through the biases they inherit from their training data and that they learn from their interactions with people. An early and infamous example of this is the chatbot Tay, developed by Microsoft. Launched in 2016, Tay was released into the habitat of Twitter and programmed to learn from its interactions with users on that platform. However, Tay had to be pulled in less than 24 hours because its encounters with people quickly taught it to be 'a racist asshole' (Vincent, 2016).

While generative AI and other algorithmic technologies are today typically trained on very large and nuanced datasets, they still inherit biases from their training data and tend to not only reproduce but also enhance these biases once implemented. Known issues include perpetuating stereotypical views of women, misgendering trans people based on binary understandings of voice and speech patterns, passing false judgement on racialized individuals and generally impacting the lives of minoritized individuals and groups negatively (Addib-Moghaddam, 2023).

To alleviate the potential harms caused by AI technologies, efforts are put into curating and cleaning training data, shifting the problem upstream to a growing precariat of annotators and other task workers, often located in the Global South, who experience the vitriol of the internet unfiltered when removing exemplars of datafied affect that are deemed unfit for algorithmic consumption (Dzieza, 2023; Perrigo, 2023). Further, guardrails are installed and algorithms are tuned, sometimes leading to issues of overcorrection, as when in early 2024 Google's image generator Gemini began depicting Nazi soldiers as people of colour (Robertson, 2024). What these attempts at problem solving that have turned out to be problematic teach us is that we are unlikely to get rid of online hate any time soon, circulating in and intensifying through nasty conflict as it does. Hence, the identification and promotion of affective alternatives becomes an all the more urgent task.

What's love got to do with it?

Love is the antonym of hate, but is it also its antidote? That seems to be Martha C. Nussbaum's (2013) position in her discussion of 'why love matters for justice'. Here, public love of one's nation (aka patriotism) is presented as a necessary ingredient of well-functioning democracies. Patriotism, says Nussbaum (2013, p 256), is not a good thing in itself, but she insists it can be non-exclusionary, bringing people together in ways that do not rely on a dichotomy of love and hate, but support open-ended relationships. Thus, she claims that patriotism does not have to involve an us–them distinction, but can sidestep hateful exclusion altogether to, instead, focus on connecting the narrative of the nation to that of the world (Nussbaum, 2013, p 214). Love in this conception is anything but the 'second-hand emotion' that Tina

Turner sings of; on the contrary, Nussbaum posits it as the primary structure of feeling for healthy democracies.

This, Nussbaum concedes, remains a substantial patriotism, one that is committed to the specific history and unique values of one's own nation rather than to abstract, procedural principles of what Habermas calls 'constitutional patriotism' (Nussbaum, 2013, p 222). Love, whether of collectives or individuals, is by definition substantial and partial; it is invested and committed. We do not love universal rules or commit to abstract regulations; we love the specifics and pledge allegiance to substantial communities. The trick, as Nussbaum sees it, is to constitute nations (or other polities) whose external boundaries are not demarcated by hate, which may be difficult, but does not make the idea of a dispassionate state any more realistic or, indeed, desirable. The question, for Nussbaum, is not how to establish love as an impartial principle, but how to practise love in a way that is inclusive and expansive, how to inspire 'strong love of a particular without coercive homogeneity or misplaced values' (Nussbaum, 2013, p 225).

The circulation of public love as a positive affective energy, one that opens up processes of meaning formation to difference and disagreement may, in principle, be possible since the type of patriotism Nussbaum envisions involves not only pride of the nation but also empathy for others and recognition of the common humanity of all. It is, we might say, both an expansive and ungreedy love that can be extended beyond existing community members while recognizing others' right to form different(ly) loving relationships. In this conception, 'love is love', as Obama said in support of marriage equality (see Chapter 5, p 103); whether at the individual level of private relationships or at the political level of public institutions, I recognize you as a love-able subject, *and* I recognize that your love is different from mine.

Nussbaum makes her argument in the institutional context of the nation state, a context that has not received much attention in this book, but nevertheless remains the anchor point of most democratic debate. Digital traversals of the boundaries of time and place notwithstanding, national borders continue to define legislative and political debate to such a degree that talk of love of the nation is not irrelevant in the context of discussing alternatives to technologically induced polarization (Bradford, 2023).

However, in its current online circulation, love bears very little resemblance to Nussbaum's lofty ideal. Rather, it is a divisive political force, bringing 'us' together in our love for the nation (or for dogs, MMA fighting or whatever is your sweet consensus) and against everyone who is not a part of our community (as sweet consensus and nasty conflict reinforce each other). As a case in point, the social rise of new nationalisms is far from incidental to but is inextricably entangled with current technological developments

(Mihelj and Jiménez-Martínez, 2021). Whatever else it does and whichever communities it supports, digitalization has also spurred new love (and hate!) for old institutions, whether tied to a nation's political structures or (as is more often the case) its cultural heritage.

Within platform infrastructures, then, love is not detached from hate; on the contrary, the two circulate as each other's co-dependent opposites, the one confirming the unity of the community to which one belongs internally and the other policing its boundaries externally (Ahmed, 2004). Thus, love and hate may spin affective circulation in different directions, but they contribute to the same overall dynamic, meaning that the rhetorical circulation of love, when posited and practiced in a violent hierarchy with hate, does not facilitate controversial encounters.

Alternatively, as Illouz says in a direct critique of Nussbaum, in democratic societies the political bond should not be lovingly bound; instead, 'it starts with difference and strangeness as well as with conflict' (2023, p 168). For Illouz, love does not fulfil any of Nussbaum's noble potentials, but becomes a source of populist pride and loyalty, building each inclusion on concerted exclusion of everything and everyone that is 'other' to the nation. In the current climate, whatever positive energies public displays of affection might potentially set in motion, the effort is in vain; paraphrasing the Bard himself, we might say that 'love's labour's lost' as long as it plays out in confrontation with hate.

In searching for alternatives to destructive circulations of love–hate relationships, we might note that algorithms are increasingly dappling in 'hot love and emotion' (to quote from Drake, a latter-day bard), as the ever-suppler interpretation of behavioural data moves technologies closer and closer to users, shifting the focus from attention and personalization to ever-deeper forms of 'intimacy capitalism' where value-creation is driven by chatbots' (and other AI technologies') ability to recognize and respond adequately to human users' feelings (Søgaard et al, 2023). While I don't want to dispute that people can – and indeed do – experience their relationships with AI companions like Replika as deep and meaningful, what is characteristic of these technologies is that they are designed to be as attentive and attuned to the needs of human interlocutors as possible.

Ultimately, then, these relationships, whatever else they might be, are not encounters between individuals with different opinions and diverse experiences, but feedback loops in which algorithms seek to respond to input in the manner that will be most pleasing to human interlocutors. In other words, when they work, these relationships reduce rather than produce friction; they may feel like meaningful encounters, but they are not controversial. They do not invite sustained interaction around differing views, but instead provide new ways of keeping people involved in one-sided conversations, of increasing individual engagement in sociotechnical relations simply because these relations feel good. As such, they may feel 'hot', but at the same time they take Illouz's (2007) 'cold intimacies' to new heights (or is it lows?), moving beyond monetizing the hook-up to make 'techno-emodities' of entire (romantic) relationships (cf. Illouz and Kotliar, 2022).

We should certainly be looking for means of circulating love independently of hate, but such circulation does not support controversial encounters if it simply takes the form of sweet consensus, which is the case for both social media's circulation of endearing affective signs (like kittens and puppies) and for encounters with AI systems that tend to be so eager to please that they are as likely to spur human irritation as to attract infatuation (Eliot, 2020).

Even if AI systems offer increasingly intimate relationships in which people might become fully invested, it continues to matter that the feelings are not mutual. As such, the 'artificial' vulnerability of AIs, the sense in which they appear to experience human emotions, is only felt at the end of human interlocutors who become vulnerable to the risk of losing a beloved companion with every system update (Verma, 2023).

Love, in sum, has nothing to do with controversy unless it is accompanied by 'riskier' feelings like passion and desire – that is, unless something is truly at stake throughout the encounter. This requires mutual vulnerability, an openness of all parties to being influenced by the other, which may lead to great common experiences, but can also cause hurt and suffering. If we do not risk anything in the encounter and if there is no tension or uncertainty involved, there is no controversy either, and hence no democratic potential.

Becoming mutually vulnerable

This leads us to the issue of affective alternatives to love and hate, asking what emotional tenors might drive affective circulations that open up spaces for debate instead of shutting them down. This question is at the heart of Mouffe's (2022) and Illouz's (2023) common search for alternatives to currently dominant populisms; whether that search is conducted from within or outside of the circuits of populist affect, the central issue is what structures of feeling may support a relationship between people who are in conflict with each other. This involves the classical issue of solidarity with strangers, of forming social bonds with people who are different from us – or, in other words, how might we come to recognize each other, thereby constituting each other as 'socially viable beings' (Butler, 2004, p 2), *with* our disagreements rather than basing recognition on the act of overcoming differences? With all the fear and anxiety it might involve, have can we give ourselves over to the radical relationality of the controversial encounter?

Butler discusses the risks involved in such relationality in terms of the inherent exclusivity of existing 'norms of recognition', pointing out how 'the demand to be recognized' incurs a crucial dilemma:

> On the one hand, living without norms of recognition results in significant suffering and forms of disenfranchisement that confound the very distinctions among psychic, cultural, and material consequences. On the other hand, the demand to be recognized, which is a very powerful political demand, can lead to new and invidious forms of social hierarchy. (Butler, 2004, p 115)

Thus, recognizing one person or collective, thereby making their lives liveable, may exclude someone else, and the question remains whether all norms of recognition are inherently exclusionary.

This is a key issue in current discussions around 'identity politics', which increasingly address the question of whether social justice has somehow gone 'too far'. Should we not, a reporter asked me recently, recognize the position of those who feel that, for instance, the recognition of the rights of LGBTQIA+ individuals is somehow a violation of *their* rights? In other words, have we swung too far in the direction of 'the other hand' in Butler's dilemma? We might (as indeed I did) counter that there still is a while to go before the rights of, say, White, cis, straight, middle-aged, middle-class, ablebodied men are truly impinged, just as we can argue that basic human rights should take precedence over the right to not have one's privilege challenged, and, further, we can insist that the comfort of consensus is not actually a right, but really a symptom of a system in which rights and privileges are unequally distributed. Still, the fact that we are having this conversation indicates that people with privilege are discovering the fragility of their position.

In Butler's words, 'the established discourse remains established only by being perpetually re-established, so it risks itself in the very repetition it requires' (2000, p 41). It is this risk that has become evident to many, leading to the backlash that is currently rolling out across a swathe of issues, from LGBTQIA+ and other minority rights to questions of climate justice and planetary responsibility. What the backlash articulates is, essentially, a concern with privilege – or, rather, with the question of how to maintain privilege when challenged by alternative discourses that are not willing to hold up the vestiges of existing hierarchies, but send new norms and different subject positions into circulation.

Rather than indicating the need for us 'to strike a balance', what the current discourse makes clear is the danger of framing the matter of the 'right to recognition' as a zero-sum game in which more recognition for some necessarily means less recognition for others. As long as we play that game, less privileged groups will be at the mercy of the more privileged, asking for their recognition and living under their norms – or in direct rebellion against them. When living in such a society, the less privileged can only insist on their right to existence, taking a stance again and again – and being removed again and again. As such, the police carrying Thunberg away from a windmill, a coalmine or the Swedish Parliament (Reuters, 2024) has become a spectacle in its own right, with Thunberg holding her body still, even smiling slightly in some of the images I have seen. But in this society (our society) the less privileged also remain vulnerable to the risk that one day the privileged will deem that now recognition has gone 'too far' and begin to retract it. One day, any day, someone might have had 'enough' and could end Thunberg's (or anyone's) vulnerable resistance.

The resistance of the vulnerable becomes especially precarious when coupled with the closing of the rhetorical mind, the disengagement of differing positions from each other. In this constellation, which, as I have argued, significantly shapes digital public debate, the risk of suffering and harm on the part of minoritized communities intensifies, as seeking to persuade others to recognize the legitimate claims of such communities becomes an increasingly suspect practice. This further aggravates the urgency of finding ways to circulate affect that enable different people to recognize each other mutually, thereby also making them mutually vulnerable. In Butler's (2020, p 24) view, what is needed are structures of feeling that support radical relationality, constituting interdependence as the foundational norm of recognition. What is needed is the mutualization of vulnerability.

Vulnerability already exists and is circulated in many forms, some of which are deeply tragic, others profoundly joyous. Exploitative relations of power make for tragedies of vulnerability, whereas vulnerability that is freely given forms the basis of joy. In its public circulation vulnerability may, similarly, spin in negative as well as positive directions. Thus, we might speculate that one reason why the images of Thunberg in her early days of school strike were so powerful was the acute vulnerability they emanated. Further, public accounts of Thunberg's neurodiversity may have exposed her to patronizing commentary, but she herself sees the condition as her 'superpower' (McNamara, 2019).

Even as Thunberg seeks to reassess her vulnerability as a strength, the resulting relationship is unidirectional, organized in a violent hierarchy that subordinates the less powerful to the more powerful, making one vulnerable to the other and not vice versa. When circulated in this manner, vulnerability re-enacts the master–bondsman relationship (see Chapter 2, p 38), making the body of the less powerful available to the more powerful, who may use it as a vessel for their own repressed vulnerability. As such, positions of power 'are effectively built through a denial of their own vulnerability. This denial or disavowal requires one to forget one's own vulnerability and project, displace, and localize it elsewhere' (Butler, Gambetti and Sabsay, 2016, p 4).

Here, however, is also the clue to changing the circulation: the less powerful may not become less vulnerable, but they can refuse to carry the vulnerability of others and, instead, display their own. They can refuse to do the emotion work of others and, instead, make emotions work for themselves – as does Thunberg when making the police carry her away, making the body of power work to remove her body, thereby revealing the relationality of their bodies, or when turning potentially hurtful comments into occasions for confrontation, often with a humorous twist (see the next section). Hence, vulnerability may become a powerful form of resistance, one that possesses vulnerable groups with agency, emanating from their very positions of vulnerability.

In sum, the affective circulation of vulnerability is radically different from that of hate – and even if it often has more affinities with love, that's not exactly how it circulates either. Vulnerability is all about becoming open – and opening things up; therefore, it is a prerequisite of change. But its potential can only be realized when everyone recognizes their mutual vulnerability. Until that happens, and to guard against the negative effects of being singly vulnerable, some protection is needed.

An armour of something

Having an armour, Brené Brown (2018) says in her bestselling book on brave leadership, insulates us from influence – most notably from the influence of others on us, but also from our chance of influencing them. Having an armour may, she argues, offer a sense of security, a sense of being in charge in an unpredictable world. However, putting on an armour does not provide us with the courage we need to engage with other people; on the contrary, it shuts us off from engagement and blocks us from forming meaningful relations. That is why, in Brown's view, courageous leadership begins with taking off the armour, with daring to expose the self to others. I agree with Brown that the armour of 'control and predict', of seeking to reduce insecurity and take charge of situations, which is so commonly donned by corporate managers and other people with power, is not productive in relation to those over which the powerful are able to wield their power. In this regard, mutual vulnerability is only possible if the people who can choose to be vulnerable actually make that choice. If and when the interpellated 'we' are in positions of power, we should indeed shed our armour.

However, I disagree that meeting the world without an armour is always the better stance – or, rather, it is a very privileged position that can only be taken by those who are already protected, who have enough security to know that showing vulnerability will indeed not kill them, but will only make them stronger. For those who cannot choose vulnerability, for those who put themselves at risk whenever they enter social relationships, having recourse to some form of armour may be essential. Without an armour, vulnerable groups and individuals risk being shot through by everything and everyone – and particularly by the force of the more powerful.

Beyond questions of power, we all, I believe, need protection when entering public debate with a mind that is open to persuasion. Without an armour of something, we simply become too vulnerable to the hurt that is going to come with the insistence on open relationships. When we engage with opinions that are different from our own, tensions are going to build, and things will get uncomfortable. Sometimes we need some protection to be able to continue engaging in and with the heat of arguments. We need an armour to make continuous disagreement bearable.

Humour is one such armour – and one that I believe holds particular potential for affectively sustaining controversial encounters on digital platforms. This may seem a strange suggestion given the many instances in which humour, typically in the form of insults and ridicule, can be a negative force in online debate, circulating in close proximity with hate and bigotry (Green, 2019; McSwiney and Sengul, 2023). However, what I find appealing about humour is that it is an inherently ambiguous force: it can be used to let steam out just as it can build up tension; it can be a mechanism of inclusion just as it can be exclusionary; it can discipline and dominate, and it can be one (sometimes the only) form of critique available to the powerless, a means of rebellion and resistance. While Dean (2019) is critical of affective circulation, 'which is most powerful when it contains different, conflicting meanings' (2019, pp 331–332), this is where I see the most potential for controversy, for meaning formation that is hospitable to and capable of sustaining difference.

This is not to say that we should be uncritical of humour; rather, its disciplining potential must be criticized as harshly as its subversive force must be encouraged. This duality is neatly illustrated by a pair of tweets: one by Donald Trump about Greta Thunberg and one by Thunberg about Trump. When, in December 2019, Thunberg was elected person of the year by *Time* magazine, Trump tweeted: 'So ridiculous. Greta must work on her Anger Management problem, then go to a good old fashioned movie with a friend! Chill, Greta, Chill!' Now, one should never explain jokes for fear of ruining them, but this one is neither funny and nor does it deserve to be preserved; rather, it is one of many instances of Trump wielding the full force of humour's disciplining power. With the explicit evaluation of the event as 'ridiculous' (a staple of Trumpian discourse), the invitation is patently not to common laughter, but is rather a dismissal of Thunberg as unworthy of further consideration, which is amplified by the suggestions that her anger is problematic and that she needs to 'chill'. In other words, Trump seeks to strip Thunberg of her agency and stop further consideration of her position, even as others recognize her stance and celebrate her impact.

In immediate response to Trump's tweet, Thunberg changed her Twitter bio to: 'A teenager working on her anger management problem. Currently chilling and watching a good old fashioned movie with a friend' (Lim, 2019a).

This, incidentally, is a move Thunberg has employed in response to several dismissive Trump tweets (see Lim, 2019b). And a year later, when Trump was trying to annul the results of the 2020 US election, Thunberg tweeted: 'So ridiculous. Donald must work on his Anger Management problem, then go to a good old fashioned movie with a friend! Chill, Donald, Chill!' I find these interventions to be genuinely funny and, hence, ask you to bear with me as I explain their workings anyway. Both are classically humorous in their intertextual repetition of Trump's tweet and their failure to make the reference explicit, which means you have to 'get it' to fully appreciate the joke. The bio, to anyone who knows Thunberg in the least and is aware of Trump's tweet, is deliciously ironic; to those not in the know, it was probably mostly confusing, but it gave a lot of people a chance to display their knowledgeability. The tweet's main mechanism of mirth is the reversal, which is not only funny, but 'funny because it's true', as the saying goes. Trump really ought to have shut up already and accepted the result, and him being told so in his own words – and by the person who he unfairly directed them at – is a splendid dig. Here, quoting back incurs a reversal in positions; the less powerful has now gained a position from which to tell the more powerful to shut up – and while having no influence on Trump, Thunberg creates an opportunity for herself and all of us to laugh at him.

Trump demonstrates how bullies use ridicule to discipline their adversaries, rehearsing tropes that are well known to and have been thoroughly deconstructed in feminist humour studies, not least those focusing on online contexts (Harlow, Rowlett and Huse, 2020). Thus, humour can be deeply problematic through its disciplining capacities, but it can also be liberating in its ability to turn the tables on who is laughing (Sundén and Paasonen, 2020). In this vein, Thunberg illustrates how even suppressive humour can become subversive, indicating the degree to which humour is ambiguous and unsettling.

Further, Thunberg's two responses to Trump's tweet demonstrate that humour can be both an armour and a sword. In the first instance, Thunberg used humour to protect herself, and in the second she used it to attack her opponent – punching up in response to his low blow. Also, humour can be disarming just as it can spur conflict. Although there is no evidence of the former in the exchange between Trump and Thunberg, we do see how humour helps sustain conflict; not between Trump and Thunberg directly, but between their two camps of followers who appear in the comments to both tweets. Thus, humour may work in tandem with the mechanisms of controversial encounters of the second kind, incentivizing 'haters' and 'followers' to keep going at each other (see Chapter 5, p 115), while the spectators cheer. Again, humour is ambiguous and, significantly, it is multifarious, which is arguably why it grabs our attention and gets us involved. By its very volatility, humour is contagious; good jokes spread well.

And while there are plenty of examples of negative (online) circulations of humour, of humour that reinforces existing views and identity positions, it also has the potential to keep controversies alive.

As for humorous circulations that close in on themselves, the internet is full of both subtle examples (that feeling when you don't quite get what everyone else is laughing at) and of crass forms of hazing and harassment when becoming part of a community is predicated on the humiliation of oneself and/or the ousting of others. However, humour can also spark open circulation; mundanely when, for instance, sharing a meme because you appreciate its creativity while not necessarily agreeing with its politics, and with higher stakes when resistance movements use humour provocatively and/or caringly, for example, through caricatures of the powers that be or as internal support mechanisms – 'we need each other so we can laugh', as Ahmed (2020) says.

Sometimes humour is a means of speaking truth to power as embodied in the classical figure of the jester (Ivie, 2002; Rossing, 2017) or performed in carnivalesque acts of parody and caricature (Hariman, 2008; Hess, 2011). In these instances humour becomes controversial and controversy humorous. As such, humour can be a powerful means of 'punching up', but it can also serve as protection from punches. It is possessed of agency even when faced with repressive adversaries, whether taking the form of spectacular acts of resistance to autocracy (for example, when the feminist-activist art collective Pussy Riot literally pisses on Putin, or at least a picture of him) or of more hidden and/or banal defiance (for example, when the Starbucks employee serves decaf to the annoying customer; see du Plessis, 2018).

Humour, then, can be what enables change, whether through out-and-out ridicule of power or by means of smaller and more insidious parodic repetitions with a difference (Just and du Plessis, 2022). But humour can also serve more conservative functions while not closing down controversy; it can be the only reason to stay connected in an otherwise intolerable environment. Similarly, humour can be a means of detachment and indifference (for example, when an action that might otherwise be seen as subversive is only performed for the 'lulz'), the engaging potential of which can always be (re-)activated (for example, when lulz shape social identities and/or turn out to be subversive forces after all) (Miltner, 2014; McDonald, 2015).

In sum, humour is good for sparking controversy and can often sustain encounters, bringing people together but not that close together, driving them apart but not that far apart. Further, humour is already one of the main affective forces of digital public debate; it is circulated and intensified across the board of digital platforms and, hence, no stranger to the algorithms that run them. If we were to infuse current dynamics of love and hate with more fun, that would not go directly against the grain of algorithmic reason, as humour tends to gain a lot of traction; people pay attention to it, engage with it and circulate it further, often even adding their own creative content. However, using humour to open up affective circulation remains a risky strategy as it may close the circuits down, becoming a negative affect, but one that is always inclined to reopening. To repeat, humour is nothing if not ambiguous – it is nothing if not controversial. Therein lies the potential encounter.

Humour may enable and sustain controversial encounters by providing an armour of something with which to endure the pressures of differences and conflict. But we can also become vulnerable to the negative forces of humour in ways that make us want to protect ourselves further rather than enable us to open up more. Thus, humour cannot stand alone; it must be accompanied by intensities of feeling that may give direction to ambiguity, revealing what motivates the reasoning. To enable different people to gain deeper recognition of each other and better understanding of each other's views, we need to set jokes aside and find a stronger common ground. Hope, I suggest, can be one such positive affective force that enables people to come together while remaining different.

I don't want your hope

In the introduction to this chapter, I posited Greta Thunberg as a potential role model for sustained encounters with difference. We have now seen how the very vulnerability of her position may make her powerful, and how she uses humour as a means of sustaining engagement even when she is under attack. Still, humour is not the only affective intensity at play in Thunberg's rhetoric – and if it were, she would hardly be as influential as she is. Rather, Thunberg also expresses anger, fear and frustration, for example, when calling out politicians and telling them she wants them to panic rather than be hopeful. And in doing so, she is anything but ambiguous. Yet I will suggest that Thunberg embodies hope even as she rejects it – that this is how she finds the strength to carry on and inspires others to do the same. Hope is what sustains controversial encounters.

In considering how hope may motivate reason, recall that from the perspective of affective intensification, persuasion is not first and foremost about specific utterances of individual actors, but about the overall rhetorical circulation of affective signs. In this model, Thunberg herself becomes an affective sign, one that is contested and therefore circulated intensely as well as openly, allowing for different opinions to not only be expressed but also to encounter each other while generally forwarding an increasing consensus as well as an increasing sense of the urgent need for action, encapsulated in the warning that 'our house is on fire'. Thus, the affective circulation of Thunberg instils hope that concerted climate action will indeed become possible before it is too late.

Here, hope may be posited as the counter force of fear; as Nussbaum asserts, 'where you fear, there too you will hope', but whereas 'hope swells outward, fear shrinks back' (2018, pp 204, 205). Hope, then, is a quintessentially positive affect, one that opens processes up and takes difference in. More fundamentally, there is a 'necessity of hope for human action; perhaps regardless of the circumstances we need the hope to act toward a future and we act because it creates hope' (Kleres and Wettergren, 2017, p 517). This, I argue, is what Thunberg has offered to millions of people: the hope that their actions matter, the belief that it is not yet too late to avert the future they fear.

Such hope, in and of itself, does not create controversial encounters. As already noted, Thunberg seems to antagonize opponents rather than make allegiances with them. However, hope does mobilize followers, and it creates momentum around Thunberg's cause, making it ever harder for decision makers to ignore. When this happens, the spectacle of controversy becomes an occasion for controversial encounters as both lovers and haters add momentum to the circulation of Thunberg's message, initiated by her basic vulnerability (the lonely and fragile figure with the handwritten sign), sustained by her (and others') humorous interventions, which pull both inward and outward at the affective circulation, fuelled by other affective forces (not least the anger that Trump sought to ridicule, and the fear and panic Thunberg seeks to imbue in world leaders) and swelling outward in ever more hopeful waves. In its totality, and beyond the voice of any one person, however influential they be, what is created here is a controversial encounter around which both conflict and consensus continue to swirl, but at the centre of which action potential grows. Thus, hope is not (necessarily) a unifying force, but one that can set the direction of continued encounters that become possible because of rather than despite remaining differences.

Is digital circulation the master's tool?

The affective circulation of Greta Thunberg intensifies love and hate, but it also offers alternatives, slits of vulnerability, humorous winks, glimmers of hope – cracks through which the light can get in, as Leonard Cohen once said (and Boygenius quoted). Beyond the question of mobilization, the issue is whether these affective alternatives amount to different structures of feeling, whether they can (re)politicize digital public debate, making disagreement good again and hence reconfiguring controversial encounters for the benefit of democracy.

In the next and final chapter, I will attempt to answer that question more fully, but here it may be useful to recall the basic framework of organizing digital communication within which this book operates. I am concerned with the sociotechnical organization of digital public debate, and with how we might organize such debate differently. Considering the economic and political powers vested in the digital organization of public opinion, it may prove impossible to harness digital technologies for democratic purposes. As we have seen, this is Dean's position:

> I remain convinced that the strongest argument for the political impact of new technologies proceeds … in the direction of post-politics. Even as globally networked communications provide tools and terrains of struggle, they make political change more difficult – and more necessary – than ever before. To this extent, politics in the sense of working to change current conditions may well require breaking with and through the fantasies attaching us to communicative capitalism. (2005, p 71)

Similarly, Mouffe (2022) does not see much potential in a digital reorganization of the postpolitical space. On the contrary, she argues that digitalization offers an escape from politics, as the 'belief that digital platforms can provide a foundation for the political order clearly chimes with the claim of third-way politicians that political antagonisms have been overcome and that left and right are "zombie categories"' (2022, p 17). Contrary to Dean, Mouffe advocates the recuperation of the political force of affect, but the two clearly agree that there is no progressive potential in digital organization.

The risk is that even as affective alternatives are articulated and circulated, for instance around Greta Thunberg's climate activism, these alternative intensifications are themselves datafied, turned into techno-emodities that fit into the logic of algorithmic reason and serve the ends of capitalism (Illouz and Kotliar, 2022). Even as alternatives emerge from within the love–hate relationship of personalized polarization, they may ultimately come to reproduce the sociotechnical configuration that is shaped by and gives shape to this structure of feeling, reproducing controversial encounters of the second kind with new intensities, but to the same old ends. Thus, we should never ignore the fact that when we are employing digital technologies, even for the circulation of affective alternatives, we are using the master's tools, the transformative potential of which are inherently limited. As Audre Lorde says, using such tools 'may allow us temporarily to beat [the master] at his own game, but they will never enable us to bring about genuine change' (1981, p 99).

Therefore, it is tempting to side with Dean and Mouffe and to simply withdraw from digital organization, seeking political potential in other spaces – or, perhaps more precisely, at other levels of abstraction. Still, if we want to change political reality, we must also recognize the current configuration of this reality. We must accept that the political has become digital, and that the digital is affective. While these two preconditions are presently configuring a postpolitical space, abandoning that space will only make matters worse.

Instead, we may begin to (re)politicize digital organization by identifying and intensifying affective alternatives. Even if the circulation of such alternatives is contained within dominant sociotechnical configurations, their very existence announces the possibility of reconfiguration. Thus, articulations of affective alternatives may not singlehandedly (re)instate controversial encounters as occasions for democratic politics, but offering different structures of feeling is a prerequisite for opening the rhetorical mind, for becoming able to engage with differently motivated reasoning. Hence, it is a starting point for 'genuine change', not just at the level of affective circulation but also in terms of its sociotechnical configuration. We can use the master's tools to demand that they change the game.

8

Make Disagreement Good Again

Are humans destroying the Earth? That's a big and sinister question, but with the polycrisis of war, AI and climate change (Soros, 2023), it is not just democracy that is at risk. Rather, we need to reckon with the threats we are posing to the entire planet (Lewis and Maslin, 2015), but to do so we must become better at discussing our common concerns in public – we must make disagreement good again. Throughout this book, I have argued that digitalization stifles public debate by turning what I have termed controversial encounters into spectacles of controversy. I will conclude by suggesting that we must find ways of (re-)encountering controversy if we are to avoid ruining everything else.

In this very important sense, the age of the Anthropocene is also the age of algorithms; we have gained great powers through technological developments, but we are not always able to wield those powers in ways that are aligned with their concomitant responsibilities. Instead of helping us act collectively to solve the grand challenges of our day and age, current sociotechnical developments tear at the fabric of democratic societies. By catering to consensus and conflict, the algorithmic organizing of publics empties out the space for controversy, turning collective meaning formation into individualized processes of detached entrenchment.

This final chapter brings together the book's full argument and offers first suggestions of how to 'make disagreement good again'. Suggesting that we (re)centre controversy, I do not advocate harmony as the solution to current predicaments, but explore how we can use divisive forces productively – how we can be different without coming apart. (Re)committing to controversial encounter, I argue, is exactly what will enable democracy to solve its other crises.

I do not in any way mean to suggest that this will be an easy process or that it is something we (you and I and people 'like us') can 'just do'. On the contrary, there are strong forces and powerful stakeholders pulling us further and further into the passive position of detached entrenchment while shifting rhetorical agency to technologies of automation. Thus, the sociotechnical configuration of controversial encounters is currently in the process of shifting from one kind to another, but at present the emerging kind seems to reinforce the problems of the configuration it is replacing. Moving from algorithmic organizing to automated persuasion only seems to increase people's suspicions of each other's voiced opinions and make them even more vulnerable to the quiet persuasion of digital technologies.

Current sociotechnical developments, then, harden the configuration of digital capitalism, embedding processes of value extraction ever deeper in reasons and motives (digitalization, that is, merges Dean's communicative capitalism and Illouz's emotional capitalism). Certainly, it will take more than a few affective alternatives (see Chapter 7) to change this dismal state of affairs. The necessary changes lie beyond individual agency; we need systemic and structural change, which is not in the power of 'ordinary' citizens and should not be their (our) responsibility. But if we are to maintain hope, we need to do something, and circulating different affective intensifications open up potential for new understandings of the current situation as well as new conceptually informed practices. As I argued in Chapter 7, processes of affective intensifications are also beyond the scope of the individual, but each articulation contributes to these processes.

Detached entrenchment through the interacting dynamics of sweet consensus and nasty conflict is the currently dominant result of algorithmic reason's organization of datafied affect into processes of personalized polarization. Yet people may use existing technological affordances in ways that lie beyond immediate invitations to action, just as the invitations themselves can be altered. Shifting digitally organized configurations of public debate will take active work with and against the organizing principles of platform infrastructures. Beginning with affect, we may work *with* digital affordances to envision alternatives; by practising more inclusive processes of circulation that sustain disagreements in tension-filled relationships, we can also begin to demand of digital technologies that they stop replacing difference with consensus or splintering disagreement into conflict.

The form of agency that is most readily invited by digital affordances is memetic circulation. Such agency cannot change anything by shear volition, but memes are powerful precisely because they are repetitions with a difference. More often than not, change will be incremental, almost imperceptible, but accepting that progress is slow going may offer us the best chance of furthering it (Villadsen, 2020). While some wait for a big bang revolution, the instant overturning of society, those who are willing to 'kiss the specifics' (see Chapter 1, p 17) can, as the feminist Marxist writer collective Gibson-Graham (1993) imply, have their revolution every day, enacting small changes with every twist and turn of existing phrases and frames, patterns and norms, while not forgetting that seizing 'the memes of production' is not the end in itself, but only a means to further – different and, perhaps, better – memetic circulation (Bristow, 2019). Hence, to seize the current moment of change and imagine emerging controversial encounters of the third kind differently is also to begin the seemingly impossible task of imagining the end of (digital) capitalism (Jameson, 1994, p xii).

The previous paragraph puts a positive spin on what has been the main argument of this book – namely, that the digital organization of communication (Chapter 2) leads to the formation of digital capitalism, which organizes digital public debate according to principles of algorithmic reason and datafied affect (Chapter 3) that invite participants to take up positions of detached entrenchment (Chapter 5) and, increasingly, make them subject to automated persuasion (Chapter 6). This process reconfigures controversial encounters (Chapter 4) in ways that impede and even disparage persuasion, closing individual and collective minds to the democratic (and democratizing) potential of rhetorical engagement with disagreement. Affective alternatives (Chapter 7) offer one starting point for changing current dynamics, and this concluding chapter will look further into potential reconfigurations.

To do so, I return to the notion of plasticity, as introduced in Chapter 1, establishing the dynamics of giving and taking (and blowing up) shape as the starting point for reconsidering digital publics as self-organizing controversies that may counter algorithmic reason, positing the body as a site of resistance to datafied affect and discussing how to interrupt the frictionless flows of digital communication. Consideration of these three alternatives (self-organizing controversies, embodied resistance and communicative interruption) leads to the question of what might be the organizing principle of the configuration that takes shape from and gives shape to them. What, in other words, might shape controversial encounters of a better kind? In response to this question, I introduce an ethics of contestability – a form of social interaction that is premised on the value of the controversial as a mode of encounter. Positing contestability as the underlying ethical principle of controversial encounters enables us to imagine the contours of a new sociotechnical configuration that neither limits reason to algorithmic procedure nor turns affect into data that can be fed to algorithms, but is instead built on reasoned motives – enabling us to encounter persuasive intent constructively. This imaginary will, I hope, leave you hopeful.

While the closing of the rhetorical mind may, at present, bar us from shaping and being shaped by the opinions of others, thereby limiting our ability to act collectively in the world, we can make disagreement good again, opening and holding space for controversial encounters and, hence, gaining the ability to unite forces without having to don the same uniform. Indeed, we may become able to maintain differences while acting together, deepening the space for legitimate decisions that exist between the extremes of conflict and consensus.

In sum, I end with a call to consider the conceptual and practical foundations for turning personalized, polarized and, increasingly, automated processes of (public) meaning formation into controversial encounters that are both real encounters and truly controversial. Conceptually, what is at stake here is a new understanding of the legitimatory potential of public debate; an understanding that places emphasis on unruly participation rather than on adherence to dominant norms and that values contestation over and above any other organizing principle. An understanding that enables us to realize the political potential in the mundane experience of connection by recognizing that we are connecting *with* disagreement (see Chapter 3, p 61). In practice, controversial encounters offer an inroad to strengthening digital democracy by enabling citizens to engage in collective pluralism and become part of plural collectives.

There is always the other option

'Technology', says Jay Brower (2018, p 51), 'is constantly altering human experience to the extent that such alteration is constitutive of what it means to be human.' We are shaped by our tools and we cannot act without them, but our tools cannot act without us either – and would not exist if we had not shaped them. Thus, the relationship between humans and technologies is neither optional nor binding, but entangled. What we do and what we make is shaped by technology, but neither action nor identity is technologically determined. Technological affordances invite a certain use, making some actions easier and more appealing and others less so, thereby inviting the formation of some configurations and neglecting alternatives.

At the moment, the invitation that shapes public debate pulls it towards automated persuasion at the collective level of the sociotechnical configuration, which corresponds to a pull towards the closing of the rhetorical mind at the level of individual agency. Thus, public debate becomes configured according to a principle of meaning formation that I have labelled the revenge of transmission theory (see Chapter 5, p 115). Within this configuration, human actors' uptake of the invitations of digital technologies to act in accordance with algorithmic reason and datafied affect shift processes of meaning formation from the overt persuasion between people who exchange opinions and towards the covert persuasion of algorithms that curate behaviours. The resulting process resembles the view of communication that is espoused by transmission theory. Here, transmission theory is both shaped by empirical reality and gives shape to this reality; like all other theories, it is an engine, not a camera (MacKenzie, 2008). The more we explain the world in a certain way, the more the world looks that way, and our current explanations and enactments of communication tend to add explanatory power and performative potential to transmission theory. Inversing Marx's 11th thesis on Feuerbach, if we want to change the world, the point is to reinterpret it.

The idea that communication is a process of transmitting messages from a sender to a receiver bears some resemblance to the theoretical engine or 'motor scheme' that Catherine Malabou (2007) identifies with the code. As a metaphor for how we think and organize, the code is like transmission; it is the turning of something into something else that can then be deciphered and turned back into its original form. A spy encodes a letter and their handler decodes it, both using an encryption device to ensure that the message is correctly encoded and decoded. An image (say, of a smiling girl in front of a burning house) is turned into a coded string of 0s and 1s (digitized) and can now be distributed as well as manipulated by anyone who has the hardware and software to turn the code back into a picture. As powerfully performative as the concept of coding is, with its risk and lure of turning us all into stochastic parrots who defer the meaning of coded patterns in our very repetition of them, it is not the only available explanation of how communication works. When we encode and decode, as Stuart Hall (1980) famously describes the communication process, this is not a matter of 'mere' transmission, but of active formation of meaning at 'both ends'. Malabou thinks of this process of giving form to meaning and meaning to form in terms of plasticity (see Chapter 1, p 14).

If we conceptualize communication as a plastic process rather than a coded one, swapping one hermeneutic motor scheme for another, we can go on to explain human actors as always already positioned within sociotechnical configurations and constituted in and through the relations that establish individual positions within these configurations. As such, we can understand communication as the meaningful constitution of relations between human and more-than-human actors. While these relations are, in turn, constitutive of the meaning we experience (as individually positioned members of a collective), they are not determined or determining (of the individual or collective); rather, they are shaped by and give shape to everyone involved – and the resulting arrangement can always be rearranged (Dahlman, Gulbrandsen and Just, 2021).

Processes of sociotechnical arrangement certainly elude any one (human) actor's intentionality, but that does not mean we have no say. Current configurations are not the result of some grand masterplan, nor will future configurations be the direct result of human volition, whether individual or collective, but that is a cause for hope rather than despair. It is a reason to continue giving shape to that which shapes us – to continue our engagement with the networks of meaning formation of which we are part and through which we ourselves become meaningful. Even if 'the other option' does not seem readily available, we can continue our search for it and, perhaps, make it appear in and through the very process of searching.

Like what you hate: curating controversial encounters

Searching for alternatives at the level of sociotechnical configurations confronts us, once again, with the question of what publics are – and what they ought to be. Publics, says Michael Warner (2002), are self-organizing, which in his account is at once a descriptive, an explanatory and a normative claim. In other words, publics exist in order to self-organize, and they exist because they self-organize – and with the very act of self-organizing, they enhance the legitimacy of the societies of which they form part. This means that a society that gives shape to and is shaped by many different publics is more legitimate than a society that is organized as – or gives shape to – only one coherent public.

This valorization of the plurality of publics is in stark contrast with the Habermasian ideal of 'the public conversation' that relies on the public sphere as a unified space for mediation between the private affairs of the citizens as left to their own devices and the public affairs of society as regulated by the state. Instead of offering legitimacy through unity, through the consensus that may (or might not actually) arise from society-wide deliberation, Warner's is a vision of legitimacy through plurality where the continued public articulation of differences – or, more precisely, the articulation of differences between publics – becomes the measure of societal wellbeing. In this sense, self-organizing publics (as was presented in more detail in Chapter 3) are the ideal shapers (and shapes) of a society that values disagreement – or is seeking to make disagreement good again. As even Dean (2003) recognizes, 'a democratic theory built around the notion of issue networks could avoid the fantasy of unity that has rendered the publicity in technoculture so profoundly depoliticizing' (2003, p 111; see also Chapter 3, p 63). As such, self-organizing around issues, understood as particular articulations of controversy, may be a source of repoliticization.

In articulating issues, self-organizing publics give shape to and take shape from controversies, defined by classical rhetoricians and modern science and technology scholars alike as the public exchange of opinions around matters of relevance to everyone in society, the complexity of which means no obvious or immediate solutions are available. We only deliberate, as Aristotle says (and I echoed in Chapter 4, p 78), about the things that are uncertain and about which we need to decide, despite being unable to resolve the uncertainties. If the truth of a matter can be calculated, there is no need to discuss it, but when matters cannot be calculated, a sustained exchange of opinions is the only means of reaching a decision that everyone can accept.

This is the foundational problem of calculated publics (Gillespie, 2014; see Chapter 3, p 54); they change the organizing principle of publics from discursive articulations to quantified relations. While this does not necessarily suppress the diverse articulation of publics, it orders publics strictly according to available articulations, tending to solidify them into 'social facts' and to restrict their ability to serve as fluid processes of social formation. Thus, Warner's notion of self-organizing publics is just as opposed to calculated publics as it is to unified ones. In fact, calculated publics are not publics at all; on the contrary, they make calculations of that which is inherently uncalculable, impeding necessary debates. Or, perhaps more precisely, they turn social organizing into primarily technological orders, shifting the balance of sociotechnical relations in the direction of depoliticized value extraction (that is, communicative capitalism).

However, beginning from the human remainder of self-organizing publics, we can work from within digital organization to curate alternative sociotechnical relationship. Thus, we may challenge automation by taking advantage of the patterns and networks it establishes, using these to circulate affective alternatives that invite different connectivities across as well as within self-organizing publics. A campaign by the Scandinavian branch of the international media organization *Vice* may provide a simple illustration of this. Entitled 'like what you hate', the campaign consisted of a simple call to engage with different opinions and offered a technical means of doing so, directing users to a site that matched each of their views of 'one side' of a matter with a view of 'the other side' (Creative Circle, 2018). While working with a binary – and, hence, limited – understanding of pros and cons, as articulated in emotions of 'liking' and 'hating', the campaign does illustrate the possibility of working through algorithmic affordances to create new and perhaps unanticipated relations.

As users become more aware of and begin negotiating their relationships with algorithms (Lomborg and Kapsch, 2020), the emotional valences of these negotiations may become more nuanced, including, for instance, irony, ambiguity and even outright resistance. Hence, self-organizing around affective alternatives may initiate a process of altering the orthodoxy of calculated publics from within.

Simply by engaging actively with algorithmic reason, individual users of digital platforms may enhance the possibility of controversial encounters, of meeting people who think differently about the matters at hand. And when doing so, agency is shifted from automated calculation and back to humans who self-organize with and around intensities of feeling. When that happens, we can establish the contours of sociotechnical self-organizing, which uses the very logics of algorithms to tune back into human articulation. Thus, we become able to shape controversies through the manner and style of our engagement.

Surely, many people will continue to engage in either nasty conflict or sweet consensus, the two role models most readily on offer. But when alternatives become available and as they begin circulating, they may become increasingly attractive. The inventory of affective alternatives, as begun in

Chapter 7, offers first pointers of intensities of feeling that may be particularly useful in this regard; vulnerability is a necessary precondition of affecting and being affected, humour may offer some protection from the hurt that being vulnerable incurs, and hope is what will enable us to go on, even when the suffering becomes unbearable and change feels impossible.

While it may feel like digital publics are overwhelmingly shaped by algorithmic reason, we can shape the outcomes of such reasoning through the input we provide. If affective alternatives gain momentum, they also gain potential to change calculated publics from within, creating new data points for algorithmic reason to identify as patterns that will, in turn, shape the calculated relations that algorithms establish between individuals as members of publics. Thus, we can begin to reconstitute publics as networks of self-organizing controversies where we encounter not only what we like and hate, but many other affective intensities of dis/agreement as well: curiosity, doubt, confusion and disorientation, to name but a few – feelings that may not be comfortable but are productive precisely because of the discomfort they incur, because of their ability to trouble existing patterns and positions.

Since feeling is first: embodying resistance

Affective responses, whether to the circulation of love, hate or subtler intensities of feeling, are first and foremost bodily. And while virtual realities may seem to complicate our relationship with physical materialities, this fact does not change: affect is an intersubjectively distributed emotion that is experienced individually (Richard and Rudnyckyj, 2009). Affect is felt before it is articulated. When we speak of our feelings, we turn bodily sensation into discursive sense. However, the process I have labelled datafied affect short circuits the intersubjective distribution of affective signs, feeding individuals with more of the feelings to which they respond most readily. This is what has led to the rise of sweet consensus and nasty conflict, respectively. And it is what may enable the establishment of affective alternatives. If we engage with alternatives, we will encounter more of them.

Conceptually, what is at stake here is another (re)turn to the body that seems to forever elude us. While the body is 'in', it is also in constant need of being brought 'back in' (Frank, 1990). As often as we disavow the mind-body split and reveal the trick of the intellectual displacement of materiality ('you be my body for me'), just as frequently do we reproduce the duality ('but do not let me know that the body you are is my body'). If we do not constantly rearticulate the materiality of the body, it tends to fall silent. As Karen Barad notes of various 'turns' in the social sciences and humanities: 'Language matters. Discourse matters. Culture matters. There is an important sense in which the only thing that does not seem to matter anymore is matter' (2003, p 801).

The challenge, as Barad poses it, is to connect matter – including, but not limited to, bodily materialities – with all the other things that matter, to understand language, discourse, culture and so on in their interrelations with materials of all sorts. Or, more precisely, to put all of these matters, social and material, on a par, showing how they are all 'vibrant', in Jane Bennett's (2010) apt terminology, and become affectively charged in and through their relations to each other (Bennett has a vivid description of her observation of a constellation involving, among other things, a glove, a stick and a dead rat in a gutter, which really drives that point home for me).

On the one hand, we may posit algorithms as just another material element in the sociotechnical configurations that involve them. On the other hand, algorithms have the ability to organize the other elements of the organizations in which they get involved. They seemingly disrupt the 'flat ontology' of sociotechnical configurations and impose their own organizing principle upon them. More specifically, algorithms complicate our conceptualization of the body once again, both hiding it from view and working on it directly. In practical terms, what I mean here is that the bodies of individual participants in online public debate are at once disconnected from and involved directly in the configuration of such debate. You can often choose to hide some or all aspects of your offline identity from other online participants, but you will nevertheless feel the affective intensity if and when your identity is invoked. Digital debate sets us free from our bodies but also holds us accountable to them – or, perhaps, more precisely, it shows us the undesirability of bracketing our (bodily) differences (Lunceford, 2018, p 143).

At the same time, algorithms organize debates by doing just that: bracketing differences – or, rather, placing each individual user within a bracket of datafied affect that is separate from the affective intensifications offered in other brackets and from the affective responses of everyone else, whether they are placed in one's own bracket or a different one. As we feel our subjective affective intensities in isolation from their intersubjective distribution, we become subject to a sort of entrenched detachment that matches what I have previously labelled detached entrenchment. We feel our own responses separately from the responses of others, deepening our detachment from them, which may in turn lead to further entrenchment. I cannot feel you and, hence, feel less accountable to your feelings – and more to my own.

Changing this process not only involves the circulation of affective alternatives, but also a recommitment to affective embodiment. Hence, we must work to bring the body back into intersubjective circulation, exposing the ways in which controversial encounters of the second kind are also disembodied encounters – and using the body to bring controversy back in. This work may be facilitated by its resemblance to playbour (see Chapter 2, p 34), enabling the depoliticized subject of digital capitalism to seize the means of prosumption for their own purposes. However, beyond this neoliberal logic of individual responsibilization (Hintz, Dencik and Wahl-Jorgensen, 2018), the playbouring subject may become collectively political through connective action (see Chapter 3, p 61). This is, for instance, the aim of the #freethenipple movement, which calls attention to gendered differences in online representations of bodies (Matich, Ashman and Parson, 2019).

Further, we can work beyond the online–offline 'divide' to expose how our bodies 'leak' data, for example, through experiments that alert users whenever applications on their digital devices generate and share data about them (Shklovski and Grönvall, 2020). By calling attention to the ways in which algorithms currently organize digital spaces by means of user data, bodily intervention (in digital spaces and beyond) can become a starting point for shifting not only the digital circulation of affect, but also the organization of this circulation. Only when we stop pretending that digital controversial encounters are not bodily, when we stop hiding that we are the bodies of such encounters, can we begin to change them more fundamentally, encountering controversies and experiencing what those controversies feel like for others.

This amounts to a re-appropriation of feeling beyond calculation, recognizing affect as lived experiences rather than data points. Thus, we become able to assert that no matter what we feel, the feelings are true in a deeper sense than that covered by the datafication of affect. Namely, as the foundation for rationality rather than impulses that can be rationally manipulated. Recognizing that 'feeling is first', as E.E. Cummings says, enables us to not only turn to the body but also to turn the body into a site of resistance, a holding ground against the algorithmic organization of meaning formation. And it attunes us to the motivated reasoning of persuasion.

I would prefer not to: interrupting flows

Engaging with affective alternatives to make new connections within the logic of algorithmic reason and centring bodily affect in order to introduce alternative engagements with persuasive processes constitute two ways in which we can begin to envision and enact controversial encounters differently, establishing social imaginaries of explicitly persuasive public debate that would allow for the possibility of persuasion but also for the option that we might 'agree to disagree' *and* continue the conversation nonetheless. These changes can, to some extent, be envisioned from within the current configuration, working with the organizing principles of algorithms and data, but we also need to reshape the processes of digital meaning formation more radically. Invoking Malabou's third sense of plasticity, we need to consider whether we should 'blow shit up' – and how we might do so.

One obvious course of action involves simply leaving digital platforms behind, a possibility that has been tested as early as 2010, when 10 May was declared 'Quit Facebook Day', and as late as March 2023, when many people began migrating to alternative platforms in response to Elon Musk's takeover of Twitter/X. But beyond going 'off the grid', which (as was mentioned in Chapter 1) is not really an option (or at least not an option that is available to everyone), what are the digital alternatives? And are any of the available options better at inviting controversial encounters?

Well yes, at the level of infrastructure platforms like Mastodon, which is open source and operates as a decentralized network, or, indeed, the Meta spinoff Threads, which is interoperable with Mastodon, already offer alternatives to the algorithmic curation of content that dominates digital debate on the larger platforms – and across the internet. However, the story of the Threads introduction may indicate some of the barriers to change: launched on 5 July 2023, the platform quickly attracted many users, peaking at 100 million only five days after the launch, but then dropping just as quickly. While the app does continue to exist, it was declared as good as dead only a month after its launch (Hummel, 2023).

Perhaps it is not surprising that a new platform owned by Mark Zuckerberg could not really establish itself as that much of an alternative to a platform owned by Elon Musk. And it is telling that, in the attempt to reassert its potential, the Thread algorithms (like those of Meta's other 'mainstream' platforms, Facebook and Instagram) have been tuned to hide rather than promote political content (Lorenz and Nix, 2024). However, if critique of ownership – or resistance to corporate social media – were the main driver of change, the migration to Mastodon, with its offer of community building, social enterprising and quality engagement (Zulli, Liu and Gehl, 2020), might have been more successful. Besides user inertia, the problem may be that 'the market' for social media has stabilized (perhaps even become saturated?), meaning that new entries are having an increasingly hard time. Despite indications that antitrust regulators are realizing it may be time to 'break up big tech' (Chee and Mukherjee, 2024) (and barring a revolution), the sociotechnical configuration of communicative capitalism is unlikely to shift fundamentally any time soon, and the most important thing we (as a society) can do is to hold the dominant tech companies responsible, demanding that they stop pretending they do not have as much power as they do.

In the absence of systemic change (or, rather, while we continue to push for it), we may also turn to and encourage the alternatives that do exist, even within the most established social media. For instance, Facebook groups offer sustained conversations that often move beyond the groups' designated topics (Thomsen et al, forthcoming). And despite its many problems of decorum, Reddit's rougher territory of internet culture, where users can up and down vote posts to topic-based communities (called subreddits), also offers fertile soil for affective self-organizing (Just and Petersen, 2023) and, perhaps, deliberation (Buozis, 2017). If we look beyond the dominant arenas of 'civic behaviour' and widen our understanding of such civic behaviour (that is, citizens' participation in public debate) to include mundane interactions of different sorts (Papacharissi, 2010, p 78; see also Chapter 3, p 62), we may find that better alternatives are available. Yet no one platform can serve as a role model for controversial encounters, nor should we envision the reconfiguration of digital public debate as being tied to one alternative. Rather, controversial encounters thrive in and as networks of networks (of networks and so on) (see Bruns, 2023).

Herman Melville's Bartleby may offer inspiration for passively resisting digital capitalism (Kang, 2021) by opting out of the free labour we perform for big tech, but 'preferring not to' get involved in digital public debate is not a democratically viable stance. On the contrary, it would exacerbate the detrimental consequences of the current configuration, as individuals detach themselves further from processes of meaning formation, becoming even more deeply entrenched spectators of combat rather than participants in competitive exchange. Since meaning formation in any meaningful sense of the concept is now fully digitalized, we must strengthen public debate from within the current conditions of possibility of digitalization.

Therefore, in the context of digital public debate, the inspiration we might take from Bartleby's logic of refusal has to do with interrupting the otherwise seamless flow of content, inserting friction into the endless feed. Users have already become preoccupied with and involved in the algorithmic curation of their feeds (Min, 2019), and they are discovering how even 'small acts of engagement' (Picone et al, 2019) may change the content they are offered and the quality of their involvement with this content. Such individual refusal to 'go with the flow' may chafe against the speed, smoothness and convenience of personalized digital experiences, creating opportunities for 'practical, affective, and emotional contestations' (Ash et al, 2018, p 1140) of the otherwise mindless flow of the infinite scroll.

As a general strategy, refusing to engage may be a creatively destructive force, but it will not offer a new shape for the configuration of digital public debate. To give such shape, we must express an alternative; we must beg to differ, thereby creating opportunities to engage differently – and to engage difference. With this final suggestion, contestation emerges as an ideal mode of engagement for controversial encounters and as the organizing principle for reconfiguring digital public debate.

I beg to differ: introducing an ethics of contestability

Throughout this book, I have argued that current sociotechnical developments lead to the closing of the rhetorical mind to the detriment of democratic societies. Also, I have suggested that to reverse this development, we must make disagreement good again, defining controversial encounters as the meeting of differently minded people who continue to engage with each other's viewpoints and arguments despite their sustained differences. As such, we should be open about our persuasive intentions and remain open to the persuasive intents of others. This amounts to an assertion of contestability as the central ethical principle for public debate, suggesting that the end of engagement should not be to reach common ground, but to continue engaging with disagreements, challenging the arguments of others and answering their challenges to our positions.

This resembles what Gunkel (2018) terms an ethics of otherness, which includes thinking about digital technologies as 'communicative Others' (Gunkel, 2009). The ethics of otherness involves a continuous broadening of our thinking about legitimate subjects:

> Although *who* counts as morally significant was something that had been initially limited to 'other men', moral thinking has evolved in such a way that it continually and necessarily questions its own restrictions and comes to encompass *what* had been previously excluded others – women, foreigners, animals, even the environment. Currently, we find ourselves standing before another fundamental challenge to this way of dividing up the world. This question – 'the machine question' – concerns the autonomous, intelligent machines of our own making, and it challenges many of the deep-seated assumptions about *who* or *what* constitutes a legitimate subject. (Gunkel, 2018b, p 26)

Treating digital technologies as Other may open up the question of whether they should have rights (Gunkel, 2018b), but more importantly (at least in this context) it opens up the opportunity for contesting the technologies, demanding that they are not only explainable, but also that they can answer to the objections that available explanations might occur. Thus, we may flatten the ontology that has been skewed by algorithmic reason, placing digital technologies within the arrangements they now seem to orchestrate. By putting all elements of any sociotechnical configuration on a par once more and asking of each node in the arrangement that it should account for – and be accountable to – its relationships, we can build better debates through and, importantly, with algorithms.

In relating with human and more-than-human actors, we must accept that while we are constituted by our relationships with others, we can never fully know the Other (be they human or not). Yet we must continue to try to make ourselves known – and to request that our interlocutors do the same. At present, digital actors are very bad at 'giving an account of themselves', as Butler (2005) might phrase the issue. The trouble is that rather than holding our technological interlocutors to human standards, we seem to be adapting human interaction to the standards of digital technologies. When algorithmic reason comes to organize meaning formation as streams of datafied affect, individuals are isolated from each other and become unable to act collectively as they are turned into data points and positioned in relation to patterns of data.

Beginning (and ending) with controversy should not in itself be controversial in pluralist democratic societies whose basic organizing principle is, ostensibly, the free exchange of different ideas and opinions. However, as we are becoming increasingly bad at having our own positions contested, we are beginning to question the very principle of contestation. The degree to which the articulation of controversy has become controversial, contested as a legitimate mode of engagement, should alert us to the depth of the current crisis of democracy. If controversy ends, democracy ends. To stop that from happening, we must beg to differ.

References

ACLU (2024) The ACLU is tracking 451 anti-LGBTQ bills in the US, https://www.aclu.org/legislative-attacks-on-lgbtq-rights.

Addib-Moghaddam, Arshin (2023) For minorities, biased AI algorithms can damage almost every part of life. *The Conversation*, https://theconversation.com/for-minorities-biased-ai-algorithms-can-damage-almost-every-part-of-life-211778.

Agamben, Giorgio (2000) *Means without Ends: Notes on Politics*. Minneapolis: University of Minnesota Press.

Aghazadeh, Sarah A., Burns, Alison, Chu, Jun, Feigenblatt, Hazel, Laribee, Elizabeth, Maynard, Lucy, Meyers, Amy L.M, O'Brien, Jessica L. and Rufus, Leah (2018) GamerGate: A case study in online harassment. In Jennifer Golbeck (ed.) *Online Harassment*. Cham: Springer, pp 179–207.

Agostinho, Daniela and Thylstrup, Nanna B. (2019) 'If truth was a woman': Leaky infrastructures and the gender politics of truth-telling. *ephemera*, 19(4): 745–775.

Ahmed, Sara (2004) Affective economies. *Social Text*, 22(2): 117–139.

Ahmed, Sara (2020) Feminists at work. *Feministkilljoys*, https://feministkilljoys.com/2020/01/10/feminists-at-work/.

Alexander, Carol (2022) After the FTX crash, here's what you need to know – The crypto bubble is already bursting. *The Guardian*, https://www.theguardian.com/commentisfree/2022/nov/23/ftx-binance-crypto-market.

Allen, Amy (2012) The public sphere: Ideology and/or ideal? *Political Theory*, 40(6): 822–829.

Anderson, James A. (1995) *An Introduction to Neural Networks*. Cambridge, MA: MIT Press.

Appel, Gil, Neelbauer, Juliana and Schweidel, David A. (2023) Generative AI has an intellectual property problem. *Harvard Business Review*, https://hbr.org/2023/04/generative-ai-has-an-intellectual-property-problem.

Aradau, Claudia and Blanke, Tobias (2022) *Algorithmic Reason: The New Government of Self and Other*. Oxford: Oxford University Press.

Arbatli, Ekim and Rosenberg, Dina (2021) United we stand, divided we rule: How political polarization erodes democracy. *Democratization*, 28(2): 285–307.

Aristotle (1928) *Politics. A Treatise on Government.* London: J M Dent & Sons Ltd.

Aristotle (1991) *On Rhetoric. A Theory of Civic Discourse.* Translated by George A. Kennedy. New York: Oxford University Press.

Arvidson, Adam (2016) Facebook and finance: On the social logic of the derivative. *Theory, Culture & Society*, 33(6): 3–23.

Ash, James, Anderson, Ben, Gordon, Rachel and Langley, Paul (2018) Digital interface design and power: Friction, threshold, transition. *Environment and Planning D: Society and Space*, 36(6): 1136–1153.

Ashcraft, Karen Lee (2022) *Wronged and Dangerous: Viral Masculinity and the Populist Pandemic.* Bristol: Bristol University Press.

Ashcraft, Karen Lee, Kuhn, Timothy R. and Cooren, François (2009) Constitutional amendments: 'Materializing' organizational communication. *Academy of Management Annals*, 3(1): 1–64.

Avelle, Michele, Di Marco, Niccolò, Etta, Gabriele, Sangiorgio, Emanuele, Alipour, Shayan, Bonetti, Anita, Alivisi, Lorenzo, Scala, Antonio, Baronchelli, Andrea, Cinelli, Matteo and Quattrociocchi, Walter (2024) Persistent interaction patterns across social media platforms and over time. *Nature*, https://doi.org/10.1038/s41586-024-07229-y.

Augustine (2022) *Augustine, On Christian Doctrine and Selected Introductory Works.* Edited by Timothy George. Nashville: B&H Academic.

Austin, John Langshaw (1962) *How to Do Things with Words.* Oxford: Oxford University Press.

Banerjee, Subhabrata Bobby (2022) Decolonizing deliberative democracy: Perspectives from below. *Journal of Business Ethics*, 181: 283–299.

Bansal, Varsha (2023) Meet the most powerful Uber driver in India. *Rest of World*, https://restofworld.org/2023/india-powerful-uber-driver/.

Barad, Karen (2003) Posthumanist performativity: Toward an understanding of how matter comes to matter. *Signs*, 28(3): 801–831.

Bastani, Aaron (2019) *Fully Automated Luxury Communism.* London: Verso.

Bateman, Tom (2021) Facebook profits off hate and that's why it won't change, says Frances Haugen. *Euronews*, https://www.euronews.com/next/2021/10/04/facebook-profits-off-hate-and-that-s-why-it-won-t-change-says-whistleblower-frances-haugen.

Baudrillard, Jacques (1994) *Simulacra and Simulation.* Ann Arbor: University of Michigan Press.

Bender, Emily M., Gebru, Timnit, McMillan-Major, Angelina and Mitchell, Margaret (2021) On the dangers of stochastic parrots: Can language models be too big? *FAccT '21: Proceedings of the 2021 ACM Conference on Fairness, Accountability, and Transparency*: 610–623.

Benjamin, Walter (1968 [1935]) The work of art in the age of mechanical reproduction. In Hannah Arendt (ed.) *Illuminations: Essays and Reflections.* New York: Schocken Books, pp 216–251.

Bennett, Jane (2010) *Vibrant Matter: A Political Ecology of Things*. Durham, NC: Duke University Press.

Bennett, W. Lance and Segerberg, Alexandra (2013) *The Logic of Connective Action: Digital Media and the Personalization of Contentious Politics*. New York: Cambridge University Press.

Beran, Dale (2019) *It Came from Something Awful: How a Toxic Troll Army Accidentally Memed Trump into Office*. New York: All Points Books.

Beyes, Timon, Chun, Wendy H. K., Clarke, Jean, Flyverbom, Mikkel and Holt, Robin (2022) Ten theses on technology and organization: Introduction to the special issue. *Organization Studies*, 43(7): 1001–1018.

Bhabha, Homi (2006) Third space. *Multitudes*, 26(3): 95–107.

Biden, Joseph R. (2021) Inaugural address, https://www.whitehouse.gov/briefing-room/speeches-remarks/2021/01/20/inaugural-address-by-president-joseph-r-biden-jr/.

Biesecker, Barbara (1992) Coming to terms with recent attempts to write women into the history of rhetoric. *Philosophy & Rhetoric*, 25(2): 140–161.

Black, Damien (2023) AI creators must be equal to artists in eyes of the law, says researcher. *Cybernews*, https://cybernews.com/editorial/ai-art-legal-rights/.

Black, Edwin (1994) Gettysburg and silence. *Quarterly Journal of Speech*, 80(1): 21–36.

Blaschke, Steffen, Schoeneborn, Dennis and Seidl, David (2012) Organizations as networks of communication episodes: Turning the network perspective inside out. *Organization Studies*, 33(7): 879–906.

Blodgett, Bridget M. (2020) Media in the post #GamerGate era: Coverage of reactionary fan anger and the terror of the privileged. *Television & New Media*, 21(2): 184–200.

Bloom, Allan (1987) *The Closing of the American Mind: How Higher Education Has Failed Democracy and Impoverished the Souls of Today's Students*. New York: Simon & Schuster.

Bogost, Ian (2007) *Persuasive Games. The Expressive Power of Videogames*. Cambridge, MA: MIT Press.

Bonini, Tiziano and Treré, Emiliano (2024) *Algorithms of Resistance. The Everyday Fight against Platform Power*. Cambridge, MA: MIT Press.

Borger, Julian (2021) Colin Powell's UN speech: A decisive moment in undermining US credibility. *The Guardian*, https://www.theguardian.com/us-news/2021/oct/18/colin-powell-un-security-council-iraq.

Bowman, Emma (2023) Security forces regain control after Bolsonaro supporters storm Brazil's Congress. *NPR*, https://www.npr.org/2023/01/08/1147757260/bolsonaro-supporters-storm-brazil-congress-lula.

boyd, danah (2011) Social network sites as networked publics: Affordances, dynamics, and implications. In Zizi Papacharissi (ed.) *Networked Self: Identity, Community, and Culture on Social Network Sites*. New York: Routledge, pp 39–58.

Bradford, Anu (2023) Whose AI revolution? *Project Syndicate*, https://www.project-syndicate.org/onpoint/ai-regulation-us-eu-china-challenges-opportunities-by-anu-bradford-2023-09?barrier=accesspaylog.

Braithwaite, Andrea (2016) It's about ethics in game journalism? Gamergaters and geek masculinity. *Social Media + Society*, https://doi.org/10.1177/2056305116672484.

Brosnan, Greg (2023) Climate activist Greta Thunberg graduates from 'school strikes'. *BBC*, https://www.bbc.com/news/science-environment-65858186.

Brown, Brené (2018) *Dare to Lead*. London: Vermilion.

Brown, Stephen, Hackley, Chris, Hunt, Shelby D., Marsh, Charles, O'Shaughnessy, Nicholas, Phillips, Barbara J., Tonks, David, Miles, Chris and Nilsson, Thomas (2018) Marketing (as) rhetoric: Paradigms, provocations, and perspectives. *Journal of Marketing Management*, 34(15): 1336–1378.

Brown, Wendy (1995) *States of Injury: Power and Freedom in Late Modernity*. Princeton: Princeton University Press.

Brown, Wendy (2015) *Undoing the Demos: Neoliberalism's Stealth Revolution*. New York: Zone Books.

Bristow, Dan (2019) Introduction. In Alfie Brown and Dan Bristow (eds) *Post Meme: Seizing the Memes of Production*. Earth, Milky Way: Punctum Books, pp 17–24.

Brower, Jay (2018) Rhetorical affects in digital media. In Aaron Hess and Amber Davisson (eds) *Theorizing Digital Rhetoric*. New York: Routledge, pp 43–54.

Bruns, Axel (2013) From presumption to produsage. In Ruth Towse and Christian Handke (eds) *Handbook on the Digital Creative Economy*. Cheltenham: Edward Elgar, pp 67–78.

Bruns, Axel (2019) *Are Filter Bubbles Real?* Cambridge: Polity Press.

Bruns, Axel (2023) From 'the' public sphere to a network of publics: Towards an empirically founded model of contemporary public communication spaces. *Communication Theory*, https://doi.org/10.1093/ct/qtad007.

Bruns, Axel and Highfield, Tim (2018) Is Habermas on Twitter? Social media and the public sphere. In Axel Bruns, Gunn Enli, Eli Skogerbø, Anders Olof Larsson and Christian Christensen (eds) *The Routledge Companion to Social Media and Politics*. New York: Routledge, pp 56–73.

Bucher, Taina (2017) The algorithmic imaginary: Exploring the ordinary affects of Facebook algorithms. *Information, Communication & Society*, 20(1): 30–44.

Bucher, Taina and Helmond, Anne (2018) The affordances of social media platforms. In Jean Burgess, Alice Marwick and Thomas Poell (eds) *The SAGE Handbook of Social Media*. London: SAGE Publications, pp 233–253.

Buozis, Michael (2017) Doxing or deliberative democracy? Evidence and digital affordances in the *Serial* subReddit. *Convergence*, 25(3): 357–373.
Burch, Sean (2018) 'Senator, we run ads': Hatch mocked for basic Facebook question to Zuckerberg. *The Wrap*, https://www.thewrap.com/senator-orrin-hatch-facebook-biz-model-zuckerberg/.
Burgess, Jean and Matamoros-Fernández, Ariadna (2016) Mapping sociocultural controversies across digital media platforms: One week of #gamergate on Twitter, YouTube, and Tumblr. *Communication Research and Practice*, 2(1): 79–96.
Burke, Kenneth (1945) *A Grammar of Motives*. Los Angeles: University of California Press.
Burke, Kenneth (1957) *The Philosophy of Literary Form: Studies in Symbolic Action*. New York: Vintage Books.
Burke, Kenneth (1966) *Language as Symbolic Action*. Berkeley: University of California Press.
Burke, Kenneth (1969) *A Rhetoric of Motives*. Berkeley: University of California Press.
Business Research Insights (2022) Gig economy market size, share, growth, and industry analysis, https://www.businessresearchinsights.com/market-reports/gig-economy-market-102503.
Butler, Judith (2000) Restaging the universal: Hegemony and the limits of formalism. In Judith Butler, Ernesto Laclau and Slavoj Žižek (eds) *Contingency, Hegemony, Universality: Contemporary Dialogues on the Left*. London: Verso, pp 11–43.
Butler, Judith (2004) *Undoing Gender*. New York: Routledge.
Butler, Judith (2005) *Giving an Account of Oneself*. New York: Fordham University Press.
Butler, Judith (2010) Performative agency. *Journal of Cultural Economy*, 3(2): 147–161.
Butler, Judith (2016a) *Frames of War: When Is Life Grievable?* London: Verso.
Butler, Judith (2016b) Rethinking vulnerability in resistance. In Judith Buther, Zeynep Gambetti and Leticia Sabsay (eds) *Vulnerability in Resistance*. Durham, NC: Duke University Press, pp 12–27.
Butler, Judith (2020) *The Force of Non-violence*. London: Verso.
Butler, Judith (2021) Why is the idea of 'gender' provoking backlash the world over? *The Guardian*, https://www.theguardian.com/us-news/commentisfree/2021/oct/23/judith-butler-gender-ideology-backlash.
Butler, Judith (2023) The compass of mourning. *London Review of Books*, https://www.lrb.co.uk/the-paper/v45/n20/judith-butler/the-compass-of-mourning.
Butler, Judith (2024) *Who's Afraid of Gender*. New York: Farrar, Straus & Giroux.

Butler, Judith and Malabou, Catherine (2011) You be my body for me: Body, shape, and plasticity in Hegel's phenomenology of spirit. In Stephen Houlgate and Michael Bauer (eds) *A Companion to Hegel*. Oxford: Wiley-Blackwell, pp 611–640.

Butler, Judith, Gambetti, Zeynep and Sabsay, Leticia (2016) Introduction. In Judith Buther, Zeynep Gambetti and Leticia Sabsay (eds) *Vulnerability in Resistance*. Durham, NC: Duke University Press, pp 1–11.

Campbell, Karlyn Kohrs (2005) Agency: Promiscuous and protean. *Communication and Critical/Cultural Studies*, 2(1): 1–19.

Campbell, Karlyn Kohrs and Jamieson, Kathleen Hall (1990) *Presidents Creating the Presidency. Deeds Done in Words*. Chicago: University of Chicago Press.

Camut, Nicolas (2023) Greta Thunberg removed by cops in Norway during anti-wind farm protest. *Politico*, https://www.politico.eu/article/greta-thunberg-detained-in-norway-during-protest/.

Capella, Joseph N. (2002) Cynicism and social trust in the new media environment. *Journal of Communication*, 52(1): 229–241.

Capella, Joseph N. and Jamieson, Kathleen Hall (1997) *Spiral of Cynicism: The Press and the Public Good*. Oxford: Oxford University Press.

Caron, James E. (2021) Satire and the public sphere: Ethics and poetics, reverse discourses, satiractivism. In Katerina Standish, Heather Devere, Adan Suazo and Rachel Rafferty (eds) *The Palgrave Handbook of Positive Peace*. Singapore: Palgrave Macmillan, pp 271–291.

Carothers, Thomas and O'Donohue, Andrew (eds) (2019) *Democracies Divided: The Global Challenge of Political Polarization*. Washington DC: Brookings Institution.

Castle, Jeremiah (2019) New fronts in the culture wars? Religion, partisanship, and polarization on religious liberty and transgender rights in the United States. *American Politics Research*, 47(3): 650–679.

Castoriadis, Cornelius (1987) *The Imaginary Institution of Society*. Cambridge, MA: MIT Press.

Center for AI Safety (2023) Statement on AI risk. https://www.safe.ai/statement-on-ai-risk#open-letter.

Chambers, Simone (2021) Truth, deliberative democracy, and the virtues of accuracy: Is fake news destroying the public sphere? *Political Studies*, 69(1): 147–163.

Chaput, Catherine (2010) Rhetorical circulation in late capitalism: Neoliberalism and the overdetermination of affective energy. *Philosophy and Rhetoric*, 43(1): 1–25.

Chaput, Catherine (2018) Neoliberalism and the rhetorical invention of counterpublic attunement. *Communication and the Public*, 3(3): 176–189.

Chayka, Kyle (2022) The age of algorithmic anxiety. *The New Yorker*, https://www.newyorker.com/culture/infinite-scroll/the-age-of-algorithmic-anxiety.

Chee, Foo Yun and Mukherjee, Supantha (2024) Google, Apple breakups on the agenda as global regulators target tech. *Reuters*, https://www.reuters.com/technology/google-apple-breakups-agenda-global-regulators-target-tech-2024-03-24/.

Chen, Gina Masullo (2017) *Nasty Talk: Online Incivility and Public Debate*. Cham: Palgrave Macmillan.

Cho, Jaeho, Ahmed, Saifuddin, Hilbert, Martin, Liu, Billy and Luu, Jonathan (2020) Do search algorithms endanger democracy? An experimental investigation of algorithm effects on political polarization. *Journal of Broadcasting & Electronic Media*, 64(2): 150–172.

Clayton, James (2022) Doubts cast over Elon Musk's Twitter bot claims. *BBC*, https://www.bbc.com/news/technology-62571733.

Cohen, Joshua and Fung, Achon (2021) Democracy and the digital public sphere. In Lucy Bernholz, Hélène Landemore and Rob Reich (eds) *Digital Technology and Democracy*. Chicago: University of Chicago Press, pp 23–61.

Cohen, Paula Marantz (2023) *Talking Cure: An Essay on the Civilizing Power of Conversation*. Princeton: Princeton University Press.

Conley, Thomas M. (1985) The virtues of controversy: *In memoriam* R. P. McKeon. *Quarterly Journal of Speech*, 71(4): 470–475.

Conley, Thomas M. (1990) *Rhetoric in the European Tradition*. Chicago: University of Chicago Press.

Conley, Thomas (2010) *Toward a Rhetoric of Insult*. Chicago: University of Chicago Press.

ContraPoints (2023) Tangent: Anti-LGBT bills. *YouTube*, https://www.youtube.com/watch?v=N26YVcSKgfUandt=1s.

Cools, Hannes, van Gorp, Baldwin and Opgenhaffen, Michael (2022) Where exactly between utopia and dystopia? A framing analysis of AI and automation in US newspapers. *Journalism*, https://doi.org/10.1177/14648849221122647.

Cooren, François (2018) Materializing communication: Making the case for a relational ontology. *Journal of Communication*, 68(2): 278–288.

Cooren, François, Kuhn, Timothy, Cornelissen, Joep P. and Clark, Timothy (2011) Communication, organizing, and organization: An overview and introduction to the special issue. *Organization Studies*, 32(9): 1149–1170.

Cooren, François, Matte, Frédérik, Benoit-Barné, Chantal and Brummans, Boris H. J. M. (2013) Communication as ventriloquism: A grounded-in-action approach to the study of organizational tensions. *Communication Monographs*, 80(3): 255–277.

Cooren, François and Taylor, James R. (1997) Organization as an effect of mediation: Redefining the link between organization and communication. *Communication Theory*, 7(3): 219–260.

Cosentino, Gabriele (2020) *Social Media and the Post-truth World Order: The Global Dynamics of Disinformation*. Cham: Springer.

Couldry, Nick and Mejias, Ulises A. (2019) Data colonialism: Rethinking big data's relation to the contemporary subject. *Television & New Media*, 20(4): 336–349.

Coupland, Douglas (2018) Douglas Coupland: 'I no longer remember my pre-internet brain'. *CNN*, https://edition.cnn.com/style/article/douglas-coupland-internet-brain/index.html.

Craig, Robert T. and Muller, Heidi L. (eds) (2007) *Theorizing Communication. Readings across Traditions*. Thousand Oaks: SAGE Publications.

Creative Circle (2018) Like what you hate, https://creativecircle.dk/arbejder/like-what-you-hate-3/.

Dahlgren, Peter (2005) The internet, public spheres, and political communication: Dispersion and deliberation. *Political Communication*, 22(2): 147–162.

Dahlgren, Peter (2018) Media, knowledge and trust: The deepening epistemic crisis of democracy. *Javnost – The Public*, 25(/1–/2): 20–27.

Dahlman, Sara, Gulbrandsen, Ib T. and Just, Sine N. (2021) Algorithms as organizational figuration: The sociotechnical arrangements of a fintech startup. *Big Data and Society*, 8(1): 1–15.

Dahlman, Sara, Just, Sine N., Petersen, Linea Munk, Lantz, Prins Marcus Valiant and Kristiansen, Nanna Würtz (2023) Datafied female health: Sociotechnical imaginaries of femtech in Danish public discourse. *MedieKultur*, 74: 105–126.

Dall, Anders (2023) Nu må politikere tale grønlandsk og færøsk i Folketinget – men de skal selv oversætte til dansk. *DR*, https://www.dr.dk/nyheder/indland/aki-talte-paa-sit-modersmaal-i-folketinget-og-markus-agter-goere-det-samme-nu-har.

Danaher, John (2019) *Automation and Utopia: Human Flourishing in a World without Work*. Cambridge, MA: Harvard University Press.

Das, Shanti (2022) Andrew Tate: Money-making scheme for fans of 'extreme misogynist' closes. *The Guardian*, https://www.theguardian.com/media/2022/aug/20/andrew-tate-money-making-scheme-for-fans-of-extreme-misogynist-closes.

Davenport, Thomas H. and Beck, John C. (2001) *The Attention Economy: Understanding the New Currency of Business*. Boston: Harvard Business School Press.

David, Derick (2023) Twitter is not dying, it's dead. *Medium*, https://medium.com/utopian/twitter-is-not-dying-its-dead-af7cd44ac236.

Davis, Ben (2023) An extremely intelligent lava lamp: Refik Anadol's A.I. art extravaganza at MOMA is fun, just don't think about it too hard. *Artnet*, https://news.artnet.com/art-world/refik-anadol-unsupervised-moma-2242329.

Davis, Jenny L. and Jurgenson, Nathan (2014) Context collapse: Theorizing context collusions and collisions. *Information, Communication & Society*, 17(4): 476–485.

Davis, Mark (2021) The online anti-public sphere. *European Journal of Cultural Studies* 24(1): 143–159.

Dean, Jodi (1999) Making (it) public. *Constellations*, 6(2): 157–166.

Dean, Jodi (2003) Why the net is not a public sphere. *Constellations*, 10(1): 95–112.

Dean, Jodi (2005) Communicative capitalism: Circulation and the foreclosure of politics. *Cultural Politics*, 1(1): 51–74.

Dean, Jodi (2009) *Democracy and Other Neoliberal Fantasies. Communicative Capitalism and Left Politics*. Durham, NC: Duke University Press.

Dean, Jodi (2019) Communicative capitalism and revolutionary form. *Millennium*. 47(3): 326–340.

De Blasio, Emiliana and Sorice, Michele (2020) The ongoing transformation of the digital public sphere: Considerations on a moving target. *Media and Communication*, 8(4): 1–5.

De Mul, Jos (2009) The work of art in the age of digital recombination. In Marianne van den Boomen et al (eds) *Digital Material: Tracing New Media in Everyday Life and Technology*. Amsterdam: Amsterdam University Press, pp 95–106.

Debord, Guy (1970) *Society of the Spectacle*. Detroit: Black & Red.

Deleuze, Gilles (2004) *Difference and Repetition*. London: Continuum.

Dencik, Lina, Hintz, Arne, Redden, Joanna and Treré, Emiliano (2019) Exploring data justice: Conceptions, applications and directions. *Information, Communication & Society*, 22(7): 873–881.

Dencik, Lina, Hintz, Arne, Redden, Joanna and Treré, Emiliano (2022) *Data Justice*. London: SAGE Publications.

Derrida, Jacques (1981) *Dissemination*. London: Athlone Press.

Derrida, Jacques (1988) *Limited Inc*. Evanston: Northwestern University Press.

Derrida, Jacques (1994) *Specters of Marx: The State of the Debt, the Work of Mourning, and the New International*. New York: Routledge.

Dewey, Caitlin (2014) The only guide to GamerGate you will ever need to read. *Washington Post*, https://www.washingtonpost.com/news/the-intersect/wp/2014/10/14/the-only-guide-to-gamergate-you-will-ever-need-to-read/.

Dhamani, N. and Engler, Maggie (2024) *Introduction to Generative AI*. Shelter Island: Manning Publications Co.

Dibbel, Julian (2005 [1993]) A rape in cyberspace. *The Village Voice*, https://www.villagevoice.com/2005/10/18/a-rape-in-cyberspace/.

Dieleman, Susan (2015) Epistemic justice and democratic legitimacy. *Hypatia*, 30(4): 794–810.

Dobusch, Leonhard and Schoeneborn, Dennis (2015) Fluidity, identity and organizationality: The communicative constitution of *Anonymous*. *Journal of Management Studies*, 52(8): 1005–1035.

Doctorow, Cory (2023) The 'enshittifcation' of TikTok. Or how, exactly, platforms die. *Wired*, https://www.wired.com/story/tiktok-platforms-cory-doctorow/.

Donawerth, Jane (1994) An annotated bibliography of the history of non-Western rhetorical theory before 1900. *Rhetoric Society Quarterly*, 24(1–2): 167–180.

Du Plessis, Erik Mygind (2018) Serving coffee with Zizek: On decaf, half-caf and real resistance at Starbucks. *ephemera*, 18(3): 551–576.

Duncan, Joe (2022) No, Elon Musk, Twitter is not the public square. *Medium*, https://joemduncan.medium.com/no-elon-musk-twitter-is-not-the-public-square-5ace4c9ec8b5.

Dwoskin, Elizabeth (2023) Come to the 'war cry party': How social media helped drive mayhem in Brazil. *Washington Post*, https://www.washingtonpost.com/technology/2023/01/08/brazil-bolsanaro-twitter-facebook/.

Dzieza, Josh (2023) AI is a lot of work. *The Verge*, https://www.theverge.com/features/23764584/ai-artificial-intelligence-data-notation-labor-scale-surge-remotasks-openai-chatbots.

Edbauer, Jenny (2005) Unframing models of public distribution: From rhetorical situation to rhetorical ecologies. *Rhetoric Society Quarterly*, 35(4): 5–24.

Eddington, Sean M. (2018) The communicative constitution of hate organizations online: A semantic network analysis of 'Make America Great Again'. *Social Media + Society*, 4(3): 1–12.

Eggers, Dave (2021) *The Every*. New York: Vintage Books.

Eika, Jonas (2019) Jonas Eikas takketale ved modtagelsen af Nordisk Råds litteraturpris, https://www.dansketaler.dk/tale/jonas-eikas-takketale-ved-modtagelsen-af-nordisk-raads-litteraturpris-2019/.

Eisikovits, Nir (2023) AI is an existential threat – just not the way you think. *The Conversation*, https://theconversation.com/ai-is-an-existential-threat-just-not-the-way-you-think-207680.

Eliot, Lance (2020) Why excessive AI politeness could be oddly inconsiderate, implications too for self-driving cars. *Forbes*, https://www.forbes.com/sites/lanceeliot/2020/08/17/why-excessive-ai-politeness-could-be-oddly-inconsiderate-implications-too-for-self-driving-cars/?sh=73115cd67523.

Enos, Richard Leo (2006) Classical rhetoric and rhetorical criticism. *Rhetoric Review*, 25(4): 361–365.

Erman, Eva (2009) What is wrong with agonistic pluralism? Reflections on conflict in democratic theory. *Philosophy and Social Criticism*, 35(9): 1039–1062.

Fazio, Marie (2021) The world knows her as 'Disaster Girl': She has just made $500,000 off the meme. *New York Times*, https://www.nytimes.com/2021/04/29/arts/disaster-girl-meme-nft.html.

Floridi, Luciano (2015) Introduction. In Luciano Floridi (ed.) *The Onlife Manifesto: Being Human in a Hyperconnected Era*. Cham: Springer, pp 1–3.

Forman, Rebecca, Shah, Soleil, Jeurissen, Patrick, Jit, Mark and Mossialos, Elias (2021) COVID-19 vaccine challenges: What have we learned so far and what remains to be done? *Health Policy*, 125(5): 553–567.

Foucault, Michel (1982) The subject and power. *Social Inquiry*, 8(4): 777–795.

Foucault, Michel (1988) *Technologies of the Self*. Amherst: University of Massachusetts Press.

Foucault, Michel (1994) *Ethics: Subjectivity and Truth*. New York: New Press.

Foucault, Michel (2001) *Fearless Speech*. Los Angeles: Semiotext(e).

Foust, Joshua and Pratt, Frankel (2021) Social media finally broke the public sphere. *Foreign Policy*, https://foreignpolicy.com/2021/01/22/social-media-broke-liberal-democracy-capitol-mob/.

Frank, Arthur W. (1990) Bringing the body back in: A decade review. *Theory, Culture & Society*, 7(1): 131–162.

Fraser, Nancy (1992) Rethinking the public sphere: A contribution to the critique of actually existing democracy. In Craig Calhoun (ed.) *Habermas and the Public Sphere*. Cambridge, MA: MIT Press, pp 109–142.

Fraser, Nancy (2022) *Cannibal Capitalism. How Our System Is Devouring Democracy, Care, and the Planet – and What We Can Do about It*. London: Verso.

Fraser, Nancy and Jaeggi, Rahel (2018) *Capitalism: A Conversation in Critical Theory*. Cambridge: Polity Press.

Frenkel, Sheera (2021) The storming of Capitol Hill was organized on social media. *New York Times*, https://www.nytimes.com/2021/01/06/us/politics/protesters-storm-capitol-hill-building.html.

Frenkel, Sheera and Conger, Kate (2022) Hate speech's rise on Twitter is unprecedented, researchers find. *New York Times*, https://www.nytimes.com/2022/12/02/technology/twitter-hate-speech.html.

Fuchs, Christian (2021) *Social Media. A Critical Introduction*, 3rd edn. London: Sage.

Future of Life Institute (2023) Pause giant AI experiments: An open letter, https://futureoflife.org/open-letter/pause-giant-ai-experiments/.

Gardiner, Becky (2018) 'It's a terrible way to go to work': What 70 million readers' comments on the Guardian revealed about hostility to women and minorities online. *Feminist Media Studies*, 18(4): 592–608.

Gaonkar, Dilip Parameshwar (1993) The idea of rhetoric in the rhetoric of science. *Southern Journal of Communication*, 58(4): 258–295.

Garsten, Bryan (2006) *Saving Persuasion: A Defense of Rhetoric and Judgement*. Cambridge, MA: Harvard University Press.

Gerbaudo, Paolo (2021) *The Great Recoil: Politics after Populism and Pandemic*. London: Verso.

Gebru, Timnit and Mitchell, Margaret (2022) We warned Google that people might believe AI was sentient: Now it's happening. *Washington Post*, https://www.washingtonpost.com/opinions/2022/06/17/google-ai-ethics-sentient-lemoine-warning/.

Geisler, Cheryl (2004) How ought we to understand rhetorical agency? *Rhetoric Society Quarterly*, 34(3): 9–17.

Geschke, Daniel, Lorenz, Jan and Holtz, Peter (2019) The triple-filter bubble: Using agent-based modelling to test a meta-theoretical framework for the emergence of filter bubbles and echo chambers. *British Journal of Social Psychology*, 58(1): 129–149.

Geyser, Werner (2022) The state of influencer marketing 2022: A benchmark report. *Influencer Marketing Hub*, https://influencermarketinghub.com/influencer-marketing-benchmark-report/.

Gibson-Graham, J.K. (1993) Waiting for the revolution, or how to smash capitalism while working at home in your spare time. *Rethinking Marxism*, 6(2): 10–24.

Giddens, Anthony (1991) *Modernity and Self-Identity*. Stanford: Stanford University Press.

Gillespie, Tarleton (2010) The politics of 'platforms'. *New Media & Society*, 12(3): 347–364.

Gillespie, Tarleton (2014) The relevance of algorithms. In Tarleton Gillespie, Pablo J. Bocskowski and Kirsten A. Frost (eds) *Media Technologies: Essays in Communication, Materiality, and Society*. Cambridge, MA: MIT Press, pp 167–193.

Gillespie, Tarleton (2018) *Custodians of the Internet: Platforms, Content Moderation, and the Hidden Decisions that Shape Social Media*. New Haven: Yale University Press.

Giridharadas, Anand (2022) *The Persuaders: At the Front Lines of the Fight for Hearts, Minds, and Democracy*. New York: Alfred A. Knopf.

Giroux, Henry A. (2022) The Nazification of American education. *Journal of Higher Education Policy and Leadership Studies*, 3(3): 7–14.

Glenn, Cheryl (1994) Sex, lies, and manuscript: Refiguring Aspasia in the history of rhetoric. *College Composition and Communication*, 45(2): 180–199.

Goode, Luke, McCullough, Alexis and O'Hare, Gelise (2011) Unruly publics and the fourth estate on YouTube. *Participations*, 8(2): 594–615.

Goodnight, G. Thomas (1982) The personal, technical, and public spheres of argument: A speculative inquiry into the art of public deliberation. *Journal of the American Forensic Association*, 18(4): 214–227.

Goodnight, G. Thomas (1999) Mssrs. Dinkins, Rangel, and savage in colloquy on the African burial ground: A companion reading. *Western Journal of Communication*, 63(4): 511–525.

Goodnight, G. Thomas (2012) The personal, technical, and public spheres: A note on 21st century critical communication inquiry. *Argumentation and Advocacy*, 48(4): 258–267.

Gorman, Amanda (2021) The hill we climb: The Amanda Gorman poem that stole the inauguration show. *The Guardian*, https://www.theguardian.com/us-news/2021/jan/20/amanda-gorman-poem-biden-inauguration-transcript?fbclid=IwAR3xMJkEep1_YQjIrZFKBClqKvTh7wNYQDwfvZmqAgDkLSFJBGPVMVtHnmY.

Graham, Megan and Elias, Jennifer (2021) How Googles $150 billion advertising business works. *CNBC*, https://www.cnbc.com/2021/05/18/how-does-google-make-money-advertising-business-breakdown-.html.

Green, Viveca S. (2019) 'Deplorable' satire: Alt-right memes, White genocide tweets, and redpilling normies. *Studies in American Humor*, 5(1): 31–69.

Greene, Ronald Water (2004) Rhetoric and capitalism: Rhetorical agency as communicative labor. *Philosophy & Rhetoric*, 37(3): 188–206.

Greenesmith, Heron (2022) Bodily autonomy under attack in the US: Access to abortion and gender-affirming care dismantled by Christian-Right theocracy. *Center for Research on Extremism*, https://www.sv.uio.no/c-rex/english/news-and-events/right-now/2022/bodily-autonomy-under-attack-in-the-us.html.

Gregg, Melissa (2011) *Work's Intimacy*. Cambridge: Polity Press.

Gregg, Melissa (2018) *Counterproductive: Time Management in the Knowledge Economy*. Durham, NC: Duke University Press.

Grieshaber, Kirsten (2023) Greta Thunberg carried away by police at German mine protest. *AP News*, https://apnews.com/article/greta-thunberg-german-mine-protest-a870ba0ba69c7816cc04f13b8be2cb94.

Gronewoller, Brian (2021) *Rhetorical Economy in Augustine's Theology*. Oxford: Oxford University Press.

Gross, Alan (2006) *Starring the Text: The Place of Rhetoric in Science Studies*. Carbondale: Southern Illinois University Press.

Gross, Alan G. (2008) Rhetoric of science. *The International Encyclopedia of Communication*, https://doi.org/10.1002/9781405186407.wbiecr075.

Gulbrandsen, Ib T. and Just, Sine N. (2011) The collaborative paradigm: Towards an invitational and participatory concept of online communication. *Media, Culture & Society*, 33(7): 1095–1108.

Gulbrandsen, Ib T. and Just, Sine N. (2024) Artificial intelligence in organizational communication: Challenges, opportunities, and implications. In Martin Ndlela (ed.) *Organizational Communication in the Digital Era: Examining the Impact of AI, Chatbots, and COVID-19*. London: Palgrave Macmillan, pp 51–77.

Gunkel, David J. (2009) Beyond mediation: Thinking the computer otherwise. *Interactions*, 1(1): 53–70.

Gunkel, David J. (2012) Communication and artificial intelligence: Opportunities and challenges for the 21st century. *Communication +1*, 1(1): 1–26.

Gunkel, David J. (2018a) Critique of digital reason. In Aaron Hess and Amber Davisson (eds) *Theorizing Digital Rhetoric*. New York: Routledge, pp 19–31.

Gunkel, David J (2018b) *Robot Rights*. Cambridge, MA: MIT Press.

Gunkel, David J. (2023) #Derrida was right (again)! *Twitter*, https://twitter.com/David_Gunkel/status/1634592409022545923.

Gunkel, David J. and Taylor, Paul A. (2014) *Heidegger and the Media*. Cambridge: Polity Press.

Gunn, Joshua and Cloud, Dana (2010) Agentic orientation as magical voluntarism. *Communication Theory*, 20(1): 50–78.

Habermas, Jürgen (1989) *The Structural Transformation of the Public Sphere*. Cambridge, MA: MIT Press.

Habermas, Jürgen (1990) *The Philosophical Discourse of Modernity*. Cambridge, MA: MIT Press.

Habermas, Jürgen (1994) Three Normative Models of Democracy. *Constellations* 1 (1): 1–10.

Habermas, Jürgen (1998a) *The Inclusion of the Other: Studies in Political Theory*. Cambridge, MA: MIT Press.

Habermas, Jürgen (1998b) *Between Facts and Norms: Contributions to a Discourse Theory of Law and Democracy*. Cambridge: Polity Press.

Habermas, Jürgen (2001a) Why Europe needs a constitution. *New Left Review*, 42(11): 5–26.

Habermas, Jürgen (2001b) *The Postnational Constellation*. Cambridge: Polity Press.

Habermas, Jürgen (2022a) Reflections and hypotheses on a further structural transformation of the political public sphere. *Theory, Culture & Society*, 39(4): 145–171.

Habermas, Jürgen (2022b) *Ein neuer Strukturwandel der Öffentlichkeit und die deliberative Politik*. Berlin: Suhrkamp.

Haimson, Oliver L. and Hoffmann, Anna Lauren (2016) Constructing and enforcing 'authentic' identity online: Facebook, real names, and non-normative identities. *First Monday*, 21(6).

Haiven, Max (2011) Finance as capital's imagination? Reimagining value and culture in an age of fictitious capital and crisis. *Social Text*, 29(3): 93–124.

Halberstam, Jack (2011) *The Queer Art of Failure*. Durham, NC: Duke University Press.

Hall, Stuart (1980) Encoding/decoding. In Stuart Hall, Dorothy Hobson, Andrew Lowe and Paul Willis (eds) *Culture, Media, Language: Working Papers in Cultural Studies 1972–79*. New York: Routledge, pp 128–138.

Hampton, Keith, Rainie, Lee, Lu, Weixu, Dwyer, Maria, Shin, Inyoung and Purcell, Kristen (2014) Social media and the 'spiral of silence'. *PewResearch Internet Project*, https://www.pewresearch.org/internet/2014/08/26/social-media-and-the-spiral-of-silence/.

Hansen, Sne Scott (2022) Public AI imaginaries: How the debate on artificial intelligence was covered in Danish newspapers and magazines 1956–2021. *Nordicom Review*, 43(1): 56–78.

Hao, Karen (2021) The Facebook whistleblower says its algorithms are dangerous. Here's why. *MIT Technology Review*, https://www.technologyreview.com/2021/10/05/1036519/facebook-whistleblower-frances-haugen-algorithms/.

Haraway, Donna (1988) Situated knowledges: The science question in feminism and the privilege of partial perspective. *Feminist Studies*, 14(3): 575–599.

Hariman, Robert (2008) Political parody and public culture. *Quarterly Journal of Speech*, 94(3): 247–272.

Harlow, Summer, Rowlett, Jerrica Ty and Huse, Laura-Kate (2020) 'Kim Davis be like…': A feminist critique of gender humor in online political memes. *Information, Communication & Society*, 23(7): 1057–1073.

Harris, Kate Lockwood and Ashcraft, Karen Lee (2023) Deferring difference no more: An (im)modest, relational plea from/through Karen Barad. *Organization Studies*, https://doi.org/10.1177/01708406231169424.

Harris, Tristan and Raskin, Aza (2023) The AI dilemma. *YouTube*, https://www.youtube.com/watch?v=xoVJKj8lcNQ.

Hauptfleisch, Wolfgang (2023) X Corp's attack on anti-hate research is … concerning. *Medium*, https://wolfhf.medium.com/x-corps-attack-on-anti-hate-researchers-is-concerning-bef7754c06f.

Hawkins, John (2023) Are NFTs really dead and buried? All signs point to 'yes'. *The Conversation*, https://theconversation.com/are-nfts-really-dead-and-buried-all-signs-point-to-yes-214145.

Hayes, Kelly (2023) Bizarre and dangerous utopian ideology has quietly taken hold of tech world. *Truthout*, https://truthout.org/audio/bizarre-and-dangerous-utopian-ideology-has-quietly-taken-hold-of-tech-world/.

Heawood, Jonathan (2018) Pseudo-public political speech: Democratic implications of the Cambridge Analytica scandal. *Information Polity*, 23: 429–434.

Heikkilä, Melissa (2023) How judges, not politicians, could dictate America's AI rules. *MIT Technology Review*, https://www.technologyreview.com/2023/07/17/1076416/judges-lawsuits-dictate-ai-rules/.

Hess, Aaron (2011) Purifying laughter: Carnivalesque self-parody as argument scheme in *The Daily Show with Jon Stewart*. In Trischa Goodnow (ed.) *The Daily Show and Rhetoric: Arguments, Issues, and Strategies*. Lanham: Lexington Books, pp 93–112.

Hintz, Arne, Dencik, Lina and Wahl-Jorgensen, Karin (2018) *Digital Citizenship in a Datafied Society*. Cambridge: Polity Press.

Hjarvard, Stig (2008) The mediatization of society: A theory of the media as agents of social and cultural change. *Nordicom Review*, 29(2): 105–134.

Hochschild, Arlie R. (1979) Emotion work, feeling rules, and social structure. *American Journal of Sociology*, 85(3): 551–575.

Hochschild, Arlie R. (1983) *The Managed Heart: Commercialization of Human Feeling*. Berkeley: University of California Press.

Hodges, Ron, Caperchione, Eugenio, van Helden, Jan, Reichard, Christoph and Sorrentino, Daniela (2022) The role of scientific expertise in COVID-19 policy-making: Evidence from four European countries. *Public Organization*, 22: 249–267.

Høj, Olivia (2023) Hun insisterede på at tale grønlandsk i folketingssalen – og så blev Karsten Hønge sur. *DR*, https://www.dr.dk/nyheder/politik/hun-insisterede-paa-tale-groenlandsk-i-folketingssalen-og-saa-blev-kars ten-hoenge.

Holtzhausen, Derina (2016) Datafication: Threat or opportunity for communication in the public sphere? *Journal of Communication Management*, 20(1): 21–36.

Houghton, Elizabeth (2019) Becoming a neoliberal subject. *ephemera*, 19(3): 615–626.

Houston, Kenneth (2018) The necessity of postmodernism in the post-truth age. *Areo*, https://areomagazine.com/2018/04/15/the-necessity-of-postmodernism-in-the-post-truth-age/.

Hsu, Hua (2023) What conversation can do for us. *The New Yorker*, https://www.newyorker.com/magazine/2023/03/20/what-conversat ion-can-do-for-us.

Huang, Kalley (2022) What is Mastodon and why are people leaving Twitter for it? *New York Times*, https://www.nytimes.com/2022/11/07/technol ogy/mastodon-twitter-elon-musk.html.

Hummel, Tyler (2023) Threads is dead – One month later. *Leaders*, https://leaders.com/news/social-media/threads-is-dead-one-month-later/.

Hutson, Matthew (2017) Why liberals aren't as tolerant as they think. *Politico*, https://www.politico.com/magazine/story/2017/05/09/why-liberals-arent-as-tolerant-as-they-think-215114/.

Illouz, Eva (2007) *Cold Intimacies: The Making of Emotional Capitalism*. Cambridge: Polity Press.

Illouz, Eva (2023) *The Emotional Life of Populism*. Cambridge: Polity Press.

Illouz, Eva and Kotliar, Dan M. (2022) Capitalist subjectivity, Tinder, and the emotionalization of the web. In Rosa Llamas and Russel Belk (eds) *The Routledge Handbook of Digital Consumption*. Abingdon: Routledge, pp 229–240.

Isager, Christine and Just, Sine N. (2005) Rhetoricians identified: A call to interdisciplinary action and how it resonated in the field of rhetoric. *Philosophy & Rhetoric*, 38(3): 248–258.

Ivie, Robert L. (2002) Rhetorical deliberation and democratic politics in the here and now. *Rhetoric & Public Affairs*, 5(2): 277–285.

Ivie, Robert L. (2007) Fighting terror by rite of redemption and reconciliation. *Rhetoric & Public Affairs*, 10(2): 221–248.

Ivie, Rovert (2015) Enabling democratic dissent. *Quarterly Journal of Speech*, 101(1): 46–59.

Jackson, Lauren Michele (2023) The invention of 'the male gaze'. *The New Yorker*, https://www.newyorker.com/books/second-read/the-invention-of-the-male-gaze.

Jameson, Fredric (1994) *The Seeds of Time*. New York: Columbia University Press.

Jamieson, Kathleen H. and Campbell, Karlyn K. (1982) Rhetorical hybrids: Fusions of generic elements. *Quarterly Journal of Speech*, 68(2): 146–157.

Jarratt, Susan and Ong, Rory (1995) Aspasia: Rhetoric, gender, and colonial ideology. In Andrea A. Lunsford (ed.) *Reclaiming Rhetorica: Women in the Rhetorical Tradition*. Pittsburgh: University of Pittsburgh Press, pp 9–24.

Jasanoff, Sheila (2015) Future imperfect: Science, technology, and the imaginations of modernity. In Sheila Jasanoff and Sang-Hyun Kim (eds) *Dreamscapes of Modernity: Sociotechnical Imaginaries and the Fabrication of Power*. Chicago: University of Chicago Press, pp 1–33.

Jasanoff, Sheila and Kim, Sang-Hyun (2009) Containing the atom: Sociotechnical imaginaries and nuclear power in the United States and South Korea. *Minerva*, 47(2): 119–146.

Jenkins, Henry (2006) *Convergence Culture: Where Old and New Media Collide*. New York: New York University Press.

Jenkins, Henry, Ford, Sam and Green, Joshua (2013) *Spreadable Media: Creating Value and Meaning in a Networked Culture*. New York: New York University Press.

Jennings, Will, Stoker, Gerry, Bunting, Hannah, Valgarðsson, Viktor Orri, Gaskell, Jennifer, Devine, Daniel, McKay, Lawrence and Mills, Melinda (2021) Lack of trust, conspiracy beliefs, and social media use predict COVID-19 vaccine hesitancy. *Vaccines*, 9(6): 1-14.

Jensen, Klaus Bruhn (2013) Definitive and sensitizing conceptualizations of mediatization. *Communication Theory*, 23(3): 203–222.

Jeong, Sarah (2023) Goodbye to all that harassment. *The Verge*, https://www.theverge.com/c/features/23997516/harassment-twitter-sarah-jeong-cancelled-social-change.

Jerrett, Adam (2022) How 'GamerGate' led the gaming industry to embrace more diverse and caring values. *The Conversation*, https://theconversation.com/how-gamergate-led-the-gaming-industry-to-embrace-more-diverse-and-caring-values-190068.

Jhaver, Shagun, Karpfen, Yoni and Antin, Judd (2018) Algorithmic anxiety and coping strategies of Airbnb hosts. *CHI '18: Proceedings of the 2018 Conference on Human Factors in Computing Systems*, https://doi.org/10.1145/3173574.3173995.

Johnson, Khari (2022) LaMDA and the sentient AI trap. *Wired*, https://www.wired.com/story/lamda-sentient-ai-bias-google-blake-lemoine/.

Jones, Charlotte and Slater, Jen (2020) The toilet debate: Stalling trans possibilities and defending 'women's protected spaces'. *Sociological Review*, 68(4): 834–851.

Jørgensen, Charlotte (1998) Public debate – An act of hostility? *Argumentation*, 12: 431–443.

Jørgensen, Charlotte (2002) The Mytilene debate: A paradigm for deliberative rhetoric. *ISSA Proceedings*, https://rozenbergquarterly.com/issa-proceedings-2002-the-mytilene-debate-a-paradigm-for-deliberative-rhetoric/.

Just, Sine N. (2005) *The Constitution of Meaning – A Meaningful Constitution?* Copenhagen: CBS PhD Series.

Just, Sine N. (2016) This is not a pipe: Rationality and affect in European public debate. *Communication and the Public*: 1(3): 276–289.

Just, Sine N. (2019) An assemblage of avatars: Digital organization as affective intensification in the GamerGate controversy. *Organization*, 26(5): 716–738.

Just, Sine N. and Berg, Kristine Marie (2016) Disastrous dialogue: Plastic productions of agency-meaning relationships. *Rhetoric Society Quarterly*, 46(1): 28–46.

Just, Sine N., Christensen, Jannick F. and Schwarzkopf, S. (2023) Disconnective action: Online activism against a corporate sponsorship at WorldPride 2021. *New Media & Society*, https://doi.org/10.1177/14614448231178775.

Just, Sine N. and du Plessis. Erik Mygind (2022) #Wegotthis: Queer parrhesia in the register of parodic paranoia. *Culture and Organization*, 28(5): 412–428.

Just, Sine N. and Petersen, Linea Munk (2023) YOLO publics: The potential for creative subversion of an online trading community. *Social Media + Society*, https://doi.org/10.1177/20563051231177953.

Just, Sine N., Storm, Kai and Bukuru, Sandra-Louise (2023) Onlife intersectionalities as flows of playbour: The case of women in gaming. *Media, Culture & Society*, 45(5): 899–915.

Kaine, Sarah and Josserand, Emmanuel (2019) The organisation and experience of work in the gig economy. *Journal of Industrial Relations*, 61(4): 479–501.

Kale, Sirin (2021) NFTs and me: Meet the people trying to sell their memes for millions. *The Guardian*, https://www.theguardian.com/technology/2021/jun/23/nfts-and-me-meet-the-people-trying-to-sell-their-memes-for-millions.

Kang, Woosung (2021) I would prefer not not-to: Critical theory after Bartleby. *Interventions*, 23(3): 356–367.

Kaun, Anne and Uldam, Julie (2018) Digital activism: After the hype. *New Media & Society*, 20(6): 2099–2106.

Kavada, Anastasia (2015) Creating the collective: Social media, the Occupy Movement and its constitution as a collective actor. *Information, Communication & Society*, 18(8): 872–886.

Kee, Joan and Kuo, Michelle (2023) Deep learning: AI, art, history, and the museum. *MoMA*, https://www.moma.org/magazine/articles/839.

Keijzer, Marijn A. and Mäs, Michael (2022) The complex link between filter bubbles and opinion polarization. *Data Science*, 5: 139–166.

Kelly, Meg and Piper, Imogen (2023) Videos of Brazil attack show striking similarities to Jan. 6. *Washington Post*, https://www.washingtonpost.com/world/2023/01/09/brazil-attack-january6-capitol/.

Kenny, Kate, Fotaki, Mariana and Vandekerckhove, Wim (2020) Whistleblower subjectivities: Organization and passionate attachment. *Organization Studies*, 41(3): 323–343.

Khachaturian, Rafael (2022) Rights without bounds: An interview with Wendy Brown. *Dissent*, https://www.dissentmagazine.org/online_articles/rights-without-bounds-wendy-brown.

Klein, Eszra (2020) *Why We're Polarized*. New York: Avid Reader Press.

Kleres, Jochen and Wettergren, Åsa (2017) Fear, hope, anger, and guilt in climate activism. *Social Movement Studies*, 16(5): 507–519.

Klinger, Ulrike and Svensson, Jakob (2020) What media logics can tell us about the internet? In Jeremy Hunsinger, Matthew M. Allen and Lisbeth Klastrup (eds) *Second International Handbook of Internet Research*. Dordrecht: Springer, pp 367–380.

Knops, Andrew (2007) Debate: Agonism as deliberation – on Mouffe's theory of democracy. *Journal of Political Philosophy*, 15(1): 115–126.

Kock, Christian (2009) Choice is not true or false: The domain of rhetorical action. *Argumentation*, 23(1): 61–80.

Kolbert, Elizabeth (2021) How politics got so polarized. *The New Yorker*, https://www.newyorker.com/magazine/2022/01/03/how-politics-got-so-polarized.

Konings, Martijn (2015) *The Emotional Logic of Capitalism: What Progressives Have Missed*. Stanford: Stanford University Press.

Korneliussen, Niviaq (2021) Niviaq Korneliussens takketale ved modtagelsen af Nordisk Råds litteraturpris, https://www.dansketaler.dk/tale/niviaq-korneliussens-takketale-ved-nordisk-raads-litteraturpris-2021/.

Kraidy, Marwan M. and Krikorian, Marina R. (2017) The revolutionary public sphere: The case of the Arab uprisings. *Communication and the Public*, 2(2): 111–119.

Kranzberg, Melvin (1986) Technology and history: 'Kranzberg's laws'. *Technology and Culture*, 27(3): 544–560.

Kücklich, Julian (2005) Precarious playbour: Modders and the digital games industry. *Fibreculture Journal*, 5.

Kvetny, Ida (2020) Lithodendrum 2020, https://kvetny.dk/virtual-reality/.

Lair, Daniel J., Sullivan, Katie and Cheney, George (2005) Marketization and the recasting of the professional self: The rhetoric and ethics of personal branding. *Management Communication Quarterly*, 18(3): 307–343.

Lanchester, John (2017) You are the product: It Zucks. *London Review of Books*, 39(16), https://www.lrb.co.uk/the-paper/v39/n16/john-lanchester/you-are-the-product

Lash, Scott (2015) Performativity or discourse? An interview with John Searle. *Theory, Culture & Society*, 32(3): 135–147.

Lazarsfeld, Paul F., Berelson, Bernard and Gaudet, Hazel (1948) *The People's Choice*, 2nd edn. New York: Columbia University Press.

Lee, Heejun and Cho, Chang-Hoan (2020) Digital advertising: Present and future prospects. *International Journal of Advertising*, 39(3): 332–341.

Lemoine, Blake (2022) Is LaMDA sentient? An interview. *Medium*, https://cajundiscordian.medium.com/is-lamda-sentient-an-interview-ea64d916d917.

Leonardi, Paul M. (2010) Digital materiality? How artifacts without matter, matter. *First Monday*, 15(6).

Lewis, Simon L. and Maslin, Mark A. (2015) Defining the Anthropocene. *Nature*, 519: 171–180.

Leys, Colin (2001) *Market-Driven Politics: Neoliberal Democracy and the Public Interest*. London: Verso.

Lilleker, Darren G. (2014) Why Twitter could be the worst kind of public sphere. *Bournemouth University*, https://news.bournemouth.ac.uk/2014/10/03/why-twitter-could-be-the-worst-kind-of-public-sphere/.

Lilleker, Darren G. (2018) Politics in a post-truth era. *International Journal of Media & Cultural Politics*, 14(3): 277–282.

Lim, Clarissa-Jan (2019a) Greta Thunberg changes her Twitter bio after Trump mocked her for being Time's person of the year. *Buzzfeed*, https://www.buzzfeednews.com/article/clarissajanlim/trump-greta-thunberg-time-person-of-the-year-2019.

Lim, Clarissa-Jan (2019b) Trump tried to troll Greta Thunberg on Twitter, and she responded in a quiet but powerful way. *Buzzfeed*, https://www.buzzfeednews.com/article/clarissajanlim/greta-thunberg-twitter-bio-trump?bfsource=relatedmanual.

Lomborg, Stine and Kapsch, Patrick Heiberg (2020) Decoding algorithms. *Media, Culture & Society*, 42(5): 745–761.

Lorde, Audre (1981) The master's tool's will never dismantle the master's house. In Cherríe Moraga and Gloria Anzaldúa (eds) *This Bridge Called My Back: Writings by Radical Women of Color*. New York: Kitchen Table: Women of Color Press, pp 98–101.

Lorenz, Taylor, Browning, Kellen and Frenkel, Sheera (2020) Tiktok teens and K-pop stans say they sank Trump rally. *New York Times*, https://www.nytimes.com/2020/06/21/style/tiktok-trump-rally-tulsa.html.

Lorenz, Taylor and Nix, Naomi (2024) Meta turns its back on politics again, angering some news creators. *Washington Post*, https://www.washingtonpost.com/technology/2024/02/10/politics-meta-threads-instagram/.

Lorenz-Spreen, Philipp, Oswald, Lisa, Lewandowsky, Stephan and Hertwig, Ralph (2023) A systematic review of worldwide causal and correlational evidence on digital media and democracy. *Nature Human Behavior*, 7: 74–101.

Lukianoff, Greg and Haidt, Jonathan (2018) *The Coddling of the American Mind: How Good Intentions and Bad Ideas Are Setting up a Generation for Failure*. New York: Penguin Random House.

Lundberg, Christian and Gunn, Joshua (2005) Ouija board, are there any communications? *Rhetoric Society Quarterly*, 35(4): 83–105.

Lunceford, Brett (2018) Where is the body in digital rhetoric? In Aaron Hess and Amber Davisson (eds) *Theorizing Digital Rhetoric*. New York: Routledge, pp 140–152.

Lunsford, Andrea A. (1995) On reclaiming Rhetorica. In Andrea A. Lunsford (ed.) *Reclaiming Rhetorica: Women in the Rhetorical Tradition*. Pittsburgh: University of Pittsburgh Press, pp 3–8.

Lupinacci, Ludmila (2021) 'Absentmindedly scrolling through nothing': Liveness and compulsory continuous connectedness in social media. *Media, Culture & Society*, 43(2): 273–290.

Luttrell, Regina, Xiao, Lu and Glass, Jon (eds) (2021) *Democracy in the Disinformation Age: Influence and Activism in American Politics*. New York: Routledge.

Maati, Ahmed, Edel, Mirjan, Saglam, Koray, Schlumberger, Oliver and Sirikupt, Chonlawit (2023) Information, doubt, and democracy: How digitization spurs democratic decay. *Democratization*, https://doi.org/10.1080/13510347.2023.2234831.

MacKenzie, Donald (2008) *An Engine, Not a Camera. How Financial Models Shape Markets*. Cambridge, MA: MIT Press.

Mahari, Robert, Fjeld, Jessica and Epstein, Ziv (2023) Generative AI is a minefield for copyright law. *The Conversation*, https://theconversation.com/generative-ai-is-a-minefield-for-copyright-law-207473.

Malabou, Catherine (2007) The end of writing? Grammatology and plasticity. *European Legacy*, 12(4): 431–441.

Malabou, Catherine (2010) *Plasticity at the Dusk of Writing: Dialectic, Destruction, Deconstruction*. New York: Columbia University Press.

Maloy, Ashley Fetters and de Vynck, Gerrit (2021) How wellness influencers are fueling the anti-vaccine movement. *Washington Post*, https://www.washingtonpost.com/technology/2021/09/12/wellness-influencers-vaccine-misinformation/.

Manzerolle, Vincent and Daubs, Michael (2021) Friction-free authenticity: Mobile social networks and transactional affordances. *Media, Culture & Society*, 43(7): 1279–1296.

Marimon, Sílvia (2022) Judith Butler: 'I am hopeful that the Russian army will lay down its arms.' *ara*, https://en.ara.cat/culture/am-hopeful-that-the-russian-army-will-lay-down-its-arms_128_4353851.html.

Marres, Noortje (2007) The issues deserve more credit: Pragmatist contributions to the study of public involvement in controversy. *Social Studies of Science*, 37(5): 759–780.

Marres, Noortje (2021) No issues without media: The changing politics of public controversy in digital societies. In Jeremy Swartz and Janet Wasko (eds) *Media: A Transdisciplinary Inquiry*. Bristol: Intellect, pp 228–243.

Marres, Noortje and Moats, David (2015) Mapping controversies with social media: The case for symmetry. *Social Media + Society*, 1(2): 1–17.

Margolin, Victor (2016) There is no there there. *Design Issues*, 32(1): 83–86.

Martinico, Giuseppe (2022) *The Tangled Complexity of the European Constitutional Process*, 2nd edn. Abingdon: Routledge.

Massumi, Brian (1995) The autonomy of affect. *Cultural Critique*, 31: 83–109.

Massumi, Brian (2014) Envisioning the virtual. In Mark Grimshaw (ed.) *The Oxford Handbook of Virtuality*. Oxford: Oxford University Press, pp 55–70.

Masullo, Gina M. and Overgaard, Christian Staal Bruun (2021) Connective democracy: A new way of thinking about deliberative democracy. *Media Ethics*, 32(2): nnp.

Matich, Margaret, Ashman, Rachel and Parsons, Elizabeth (2019) #Freethenipple: Digital activism and embodiment in the contemporary feminist movement. *Consumption Markets & Culture*, 22(4): 337–362.

Mazzoleni, Gianpietro (2015) Towards an inclusive digital public sphere. In Stephen Coleman, Giles Moss and Katy Parry (eds) *Can the Media Serve Democracy?* London: Palgrave Macmillan, pp 174–183.

McCoy, Jennifer, Rahman, Tahmina and Somer, Murat (2018) Polarization and the global crisis of democracy: Common patterns, dynamics, and pernicious consequences for democratic polities. *American Behavioral Scientist*, 62(1): 16–42.

McDonald, Kevin (2015) From Indymedia to Anonymous: Rethinking action and identity in digital cultures. *Information, Communication & Society*, 18(8): 968–982.

McNamara, Britney (2019) Greta Thunberg called autism her 'superpower' in post against haters. *Teen Vogue*, https://www.teenvogue.com/story/greta-thunberg-called-autism-her-superpower-in-post-against-haters.

McSwiney, Jordan and Sengul, Kurt (2023) Humor, ridicule, and the far right: Mainstreaming exclusion through online animation. *Television & New Media*, https://doi.org/10.1177/15274764231213816.

Mejias, Ulises A. (2010) 'Playbor' on the internet. *Afterimage*, 37(4): 2.

Mejias, Ulises A. and Couldry, Nick (2019) Datafication. *Internet Policy Review*, 8(4): 1–10.

Mello, Patrick A. (2017) Democratic peace theory. In Paul Joseph (ed.) *The SAGE Encyclopedia of War: Social Science Perspectives*. Thousand Oaks: SAGE Publications, pp 472–476.

Mendelson, Michael (2002) *Many Sides: A Protagorean Approach to the Theory, Practice, and Pedagogy of Argument*. Dordrecht: Springer.

Menking, Amanda and Erickson, Ingrid (2015) The heart work of Wikipedia: Gendered, emotional labor in the world's largest online encyclopedia. *CHI '15: Proceedings of the 33rd Annual ACM Conference on Human Factors in Computing Systems*, pp 207–210.

Metz, Rachel (2022) No, Google's AI is not sentient. *CNN Business*, https://edition.cnn.com/2022/06/13/tech/google-ai-not-sentient/index.html.

Meyer, John C. (2000) Humour as a double-edged sword: Four functions of humour in communication. *Communication Theory*, 10(3): 310–331.

Meyrowitz, Joshua (1985) *No Sense of Place: The Impact of Electronic Media on Social Behavior*. New York: Oxford University Press.

Miao, Hannah (2021) 'It's just Bernie being Bernie': How a photo of Sanders wearing mittens at Inauguration Day went viral. *CNBC*, https://www.cnbc.com/2021/01/23/bernie-sanders-inauguration-meme-heres-the-story-behind-the-photo.html.

Mihelj, Sabina and Jiménez-Martínez, César (2021) Digital nationalism: Understanding the role of digital media in the rise of 'new' nationalism. *Nations and Nationalism*, 27(2): 331–346.

Miller, Carl (2023) Antisemitism on Twitter has more than doubled since Elon Musk took over the platform – New research. *The Conversation*, https://theconversation.com/antisemitism-on-twitter-has-more-than-doubled-since-elon-musk-took-over-the-platform-new-research-201830.

Miller, Carolyn R. (1984) Genre as social action. *Quarterly Journal of Speech*, 70(2): 151–167.

Miller, Carolyn R. (2005) Risk, controversy, and rhetoric: Response to Goodnight. *Argumentation and Advocacy*, 42(1): 34–37.

Miller, Carolyn R. (2007) What can automation tell us about agency? *Rhetoric Society Quarterly*, 37(2): 137–157.

Miller, Susan (2008) *Trust in Texts: A Different History of Rhetoric*. Carbondale: Southern Illinois University Press.

Miltner, Kate M. (2014) 'There's no place for lulz on LOLCats': The role of genre, gender, and group identity in the interpretation and enjoyment of an Internet meme. *First Monday*, 19(8), https://doi.org/10.5210/fm.v19i8.5391.

Min, Seong Jae (2019) From algorithmic disengagement to algorithmic activism: Charting social media users' responses to news filtering algorithms. *Telematics and Informatics*, 43, https://doi.org/10.1016/j.tele.2019.101251.

Moisescu, Cristiana, Croffey, Amy and Bennett, Vicky (2023) Andrew Tate indicted on human trafficking and rape charges in Romania. *CNN*, https://edition.cnn.com/2023/06/20/europe/andrew-tate-charges-trial-intl-gbr/index.html.

Molleindustria (2019) The McDonald's Videogame, https://molleindustria.org/mcdonalds/.

MoMA (2023) Refik Anadol Unsupervised, https://www.moma.org/calendar/exhibitions/5535.

Morse, Jack (2022) Your privacy is at risk now that *Roe v. Wade* has fallen, experts warn. *Mashable*, https://mashable.com/article/supreme-court-roe-wade-digital-privacy.

Mouffe, Chantal (1999) Deliberative democracy or agonistic pluralism? *Social Research*, 66(3): 745–758.

Mouffe, Chantal (2013) *Agonistics: Thinking the World Politically*. London: Verso.

Mouffe, Chantal (2017) Democracy as agonistic pluralism. In Elizabeth Deeds Ermarth (ed.) *Rewriting Democracy: Cultural Politics in Postmodernity*. Abingdon: Routledge, pp 35–45.

Mouffe, Chantal (2022) *Towards a Green Democratic Revolution: Left Populism and the Power of Affects*. London: Verso.

Mohanty, Chandra Talpade (1984) Under Western eyes: Feminist scholarship and colonial discourse. *boundary 2*, 12(3)-13(1): 333–358.

Murphy, James Jerome (1974) *Rhetoric in the Middle Ages: A History of Rhetorical Theory from Saint Augustine to the Renaissance*. Berkeley: University of California Press.

Murphy, Matt (2022) China's protests: Blank paper becomes the symbol of rare demonstrations. *BBC*, https://www.bbc.com/news/world-asia-china-63778871.

Musk, Elon (2022) Given that Twitter serves as the defacto town square… *Twitter [now X]*, https://twitter.com/elonmusk/status/1507777261654605828?lang=da.

Nadim, Marjan and Fladmoe, Audun (2021) Silencing women? Gender and online harassment. *Social Science Computer Review*, 39(2): 245–258.

Nagle, Angela (2017) *Kill All Normies: Online Culture Wars from 4Chand and Tumblr to Trump and the Alt-Right*. Winchester: Zero Books.

Neff, Gina (2015) *Venture Labor: Work and the Burden of Risk in Innovative Industries*. Cambridge, MA: MIT Press.

Nelson, Maggie (2021) *On Freedom: Four Songs of Care and Constraint*. London: Jonathan Cape.

New York Times (2020) How statues are falling around the world, https://www.nytimes.com/2020/06/24/us/confederate-statues-photos.html.

Nielsen, Rasmus Kleis and Graves, Lucas (2017) 'News you don't believe': Audience perspectives on fake news. *Reuters Institute*, https://ora.ox.ac.uk/objects/uuid:6eff4d14-bc72-404d-b78a-4c2573459ab8/files/m2aa4c90ac64a0a6c7c6e10baec5953a0.

Noelle-Neumann, Elisabeth (1974) The spiral of silence: A theory of public opinion. *Journal of Communication*, 24(2): 43–51.

NPR Staff (2019) Transcript: Greta Thunberg's speech at the U.N. climate action summit. *NPR*, https://www.npr.org/2019/09/23/763452863/transcript-greta-thunbergs-speech-at-the-u-n-climate-action-summit.

Nussbaum, Martha C. (2013) *Political Emotions: Why Love Matters for Justice*. Cambridge, MA: Belknap Press of Harvard University Press.

Nussbaum, Martha C. (2018) *The Monarchy of Fear: A Philosopher Looks at Our Political Crisis*. New York: Simon & Schuster.

ODSC (2022) AI-generated art wins contest and stirs controvery online. *Medium*, https://odsc.medium.com/ai-generated-art-wins-contest-and-stirs-controversy-online-9d4616a8380a.

Olson, Kathryn M. and Goodnight, G. Thomas (1994) Entanglements of consumption, cruelty, privacy, and fashion: The social controversy over fur. *Quarterly Journal of Speech*, 80(3): 249–276.

Oltermann, Philip (2022) German thinkers' war of words over Ukraine exposes generational divide. *The Guardian*, https://www.theguardian.com/world/2022/may/06/german-thinkers-war-of-words-over-ukraine-exposes-generational-divide.

Ono, Kent A. and Sloop, John M. (1999) Critical rhetorics of controversy. *Western Journal of Communication*, 63(4): 526–538.

Oratz, Lisa T., Robbins, Tyler and West, D. Sean (2023) Recent rulings in AI copyright lawsuits shed some light, but leave many questions. *Perkinscoie*, https://www.perkinscoie.com/en/news-insights/recent-rulings-in-ai-copyright-lawsuits-shed-some-light-but-leave-many-questions.html.

Orgad, Liav (2010) The preamble in constitutional interpretation. *International Journal of Constitutional Law*, 8(4): 714–738.

Orlikowski, Wanda (2007) Sociomaterial practices: Exploring technology at work. Organization Studies, 28(9): 1435–1448.

Ortiz, Erik (2019) Teen climate activist Greta Thunberg tells Congress: 'Unite behind the science'. *NBS News*, https://www.nbcnews.com/science/environment/climate-activist-greta-thunberg-tells-congress-unite-behind-science-n1055851.

Pace, Jonathan (2018) The concept of digital capitalism. *Communication Theory*, 28(3): 254–269.

Palley, Thomas I. (2013) *Financialization: The Economics of Financial Capital Domination*. New York: Palgrave Macmillan.

Papacharissi, Zizi (2010) *A Private Sphere: Democracy in a Digital Age*. Cambridge: Polity Press.

Papacharissi, Zizi (2014) *Affective Publics: Sentiment, Technology, and Politics*. Oxford: Oxford University Press.

Pariser, Eli (2022) Musk's Twitter will not be the town square the world needs. *Wired*, https://www.wired.com/story/elon-musk-twitter-town-square/.

Pasquale, Frank (2016) *Black Box Society: The Secret Algorithms That Control Money and Information*. Cambridge, MA: Harvard University Press.

Patberg, Markus (2023) Digital fragmentation: Habermas on the new structural transformation of the public sphere. *RevDem*, https://revdem.ceu.edu/2023/02/02/digital-fragmentation-habermas-on-the-new-structural-transformation-of-the-public-sphere/.

Pearson, Jordan (2023) Elon Musk finally broke Twitter. *Vice*, https://www.vice.com/en/article/k7bm7m/elon-musk-finally-broke-twitter.

Perrigo, Billy (2023) Exclusive: OpenAI used Kenyan workers on less than $2 per hour to make ChatGPT less toxic. *Time*, https://time.com/6247678/openai-chatgpt-kenya-workers/.

Peters, John Durham (1999) *Speaking into the Air: A History of the Idea of Communication*. Chicago: University of Chicago Press.

Pew Research Center (2022) As partisan hostility grows, signs of frustration with the two-party system, https://www.pewresearch.org/politics/2022/08/09/as-partisan-hostility-grows-signs-of-frustration-with-the-two-party-system/.

Phan, Thao (2017) The materiality of the digital and the gendered voice of Siri. *Transformations*, 29: 23–33.

Phillips, Kendall R. (1999) A rhetoric of controversy. *Western Journal of Communication*, 63(4): 488–510.

Phillips, Kendall R. (2006) Rhetorical maneuvers: Subjectivity, power, and resistance. *Philosophy & Rhetoric*, 39(4): 310–332.

Phillips, Whitney (2015) *This Is Why We Can't Have Nice Things: Mapping the Relationship Between Online Trolling and Mainstream Culture*. Cambridge, MA: MIT Press.

Picone, Ike, Kleut, Jelena, Pavlíčková, Tereza, Romic, Bojana, Hartley, Jannie Møller and de Ridder, Sander (2019) Small acts of engagement: Reconnecting productive audience practices with everyday agency. *New Media & Society*, 21(9): 2010–2028.

Plato (1952) *Plato's Phaedrus*. Translated by R. Hackforth. Cambridge: Cambridge University Press.

Popper, Karl (1994 [1945]) *The Open Society and Its Enemies*. Princeton: Princeton University Press.

Posard, Marek N. (2022) Elon Musk may have a point about bots on Twitter. *Rand Corporation*, https://www.rand.org/blog/2022/09/elon-musk-may-have-a-point-about-bots-on-twitter.html.

Prakash, Thomas (2023) Holland undskylder til slaver, men Danmark har ikke gjort det: 'Skal vi så også undskylde for vikingetiden?' *DR*, https://www.dr.dk/nyheder/indland/holland-undskylder-til-slaver-men-danmark-har-ikke-gjort-det-skal-vi-saa-ogsaa.

Quinn, Kelly and Papacharissi, Zizi (2018) Our networked selves: Personal connection and relational maintenance in social media use. In Jean Burgess, Alice Marwick and Thomas Poell (eds) *The SAGE Handbook of Social Media*. London: SAGE Publications, pp 353–371.

Ramirez, Rachel (2024) What is 'new denial'? An alarming wave of climate misinformation is spreading on YouTube, watchdog says. *CNN*, https://edition.cnn.com/2024/01/16/climate/climate-denial-misinformation-youtube/index.html.

Rancière, Jacques (2004) *Disagreement: Politics and Philosophy*. Minneapolis: University of Minnesota Press.

Rand, Erin (2008) An inflammatory fag and a queer form: Larry Kramer, polemics, and rhetorical agency. *Quarterly Journal of Speech*, 94(3): 297–319.

Rasmussen, David (2019) Ideal speech situation. In Amy Alken and Eduordo Mendieta (eds) *The Cambridge Habermas Lexicon*. Cambridge: Cambridge University Press, pp 182–184.

Remnick, David (2023) What we talk about when we talk about trans rights. *The New Yorker*, https://www.newyorker.com/news/the-new-yorker-interview/what-we-talk-about-when-we-talk-about-trans-rights.

Reuters (2024) Swedish police forcibly removes Greta Thunberg from parliament entrance. *The Guardian*, https://www.theguardian.com/world/2024/mar/12/swedish-police-forcibly-remove-greta-thunberg-parliament-climate-protest.

Reyman, Jessica (2018) The rhetorical agency of algorithms. In Aaron Hess and Amber Davisson (eds) *Theorizing Digital Rhetoric*. New York: Routledge, pp 112–125.

Richard, Analiese and Rudnyckyj, Daromir (2009) Economies of affect. *Journal of the Royal Anthropological Institute*, 15(1): 57–77.

Richter, Jacob D. (2021) Writing with Reddiquette: Networked agonism and structured deliberation in networked communities. *Computers and Composition*, 59: 1–20.

Riding, Alan (2001) Grandpa Picasso: Terribly famous, not terribly nice. *New York Times*, https://www.nytimes.com/2001/11/24/books/grandpa-picasso-terribly-famous-not-terribly-nice.html.

Ritchie, David (2009) 'ARGUMENT IS WAR': Or is it a game of chess? Multiple meanings in the analysis of implicit metaphors. *Metaphor and Symbol*, 18(2): 125–146.

Ritzau (2019) Mette Frederiksen er helt uenig i prisvinders racismekritik. *DR*, https://www.dr.dk/nyheder/politik/mette-frederiksen-er-helt-uenig-i-prisvinders-racismekritik.

Robertson, Adi (2024) Google apologizes for 'missing the mark' after Gemini generated racially diverse Nazis. *The Verge*, https://www.theverge.com/2024/2/21/24079371/google-ai-gemini-generative-inaccurate-historical.

Rogers, Richard and Giorgi, Julia (2023) What is a meme, technically speaking. *Information, Communication & Society*, https://doi.org/10.1080/1369118X.2023.2174790.

Romano, Aja (2021) What we still haven't learned from Gamergate. *Vox*, https://www.vox.com/culture/2020/1/20/20808875/gamergate-lessons-cultural-impact-changes-harassment-laws.

Romano, Aja (2023) Is J.K. Rowling transphobic? Let's let her speak for herself. *Vox*, https://www.vox.com/culture/23622610/jk-rowling-transphobic-statements-timeline-history-controversy.

Roose, Kevin (2019) The making of a YouTube radical. *New York Times*, https://www.nytimes.com/interactive/2019/06/08/technology/youtube-radical.html.

Rosenblat, Alex (2018) *Uberland: How Algorithms Are Rewriting the Rules of Work*. Oakland: University of California Press.

Rossing, Jonathan P. (2016) A sense of humor for civic life: Toward a strong defense of humor. *Studies in American Humor*, 2(1): 1–21.

Rossing, Jonathan P. (2017) No joke: Silent jesters and comedic refusals. *Rhetoric & Public Affairs*, 20(3): 545–556.

Roth, Emma (2023a) Elon Musk says bots with 'good content' can use Twitter API for free. *The Verge*, https://www.theverge.com/2023/2/5/23586577/elon-musk-bots-good-content-twitters-api-free.

Roth, Emma (2023b) Twitter is just showing everyone all of Elon Musk's tweets now. *The Verge*, https://www.theverge.com/2023/2/13/23598514/twitter-algorithm-elon-musk-tweets.

Royster, Jacqueline Jones (2003) Disciplinary landscaping, or contemporary challenges in the history of rhetoric. *Philosophy and Rhetoric*, 36(2): 148–167.

Sadowski, Jathan (2019) When data is capital: Datafication, accumulation, and extraction. *Big Data & Society*, 6(1): 1–12.

Saikia, Bandana and Doshi, Kosha (2022) Rethinking explicit consent and intimate data collection: The looming digital privacy concern with *Roe v. Wade* overturned. *LSE*, https://blogs.lse.ac.uk/humanrights/2022/12/14/rethinking-explicit-consent-and-intimate-data-collection-the-looming-digital-privacy-concern-with-roe-v-wade-overturned/.

Santos, Tiago, Louçã, Jorge and Coelho, Helder (2019) The digital transformation of the public sphere. *Systems Research and Behavioral Science*, 36: 778–788.

Sartori, Laura and Bocca, Giulia (2022) Minding the gap(s): Public perceptions of AI and socio-technical imaginaries. *AI & Society*, 38: 1–16.

Savolainen, Laura (2022) The shadow banning controversy: Perceived governance and algorithmic folklore. *Media, Culture & Society*, 44(6): 1091–1109.

Saxonhouse, Arlene W. (2005) Another Antigone: The emergence of the female political actor in Euripedes' *Phoenician Women. Political Theory*, 33(4): 472–494.

Scheiber, Noam and Weise, Karen (2022) Amazon labor union, with renewed momentum, faces next test. *New York Times*, https://www.nytimes.com/2022/10/11/business/economy/amazon-labor-union.html.

Schoeneborn, Dennis, Blaschke, Steffen, Cooren, François, McPhee, Robert D., Seidl, David and Taylor, James R. (2014) The three schools of CCO thinking: Interactive dialogue and systematic comparison. *Management Communication Quarterly*, 28(2): 285–316.

Scholz, Trebor (2013) Introduction: Why does digital labor matter now? In Trebor Scholz (ed.) *Digital Labor: The Internet as Playground and Factory*. New York: Routledge, pp 1–9.

Scholz, Trebor (2016) *Platform Cooperativism: Challenging the Corporate Sharing Economy*. New York: Rosa Luxemborg Stiftung.

Schmidt, Gudrun Marie (2022) 'Vi er nødt til at stoppe det her sammen': Unge i Grønland har fået nok af landets mange selvmord. *Politiken*, https://politiken.dk/indland/art9074364/Unge-i-Grønland-har-fået-nok-af-landets-mange-selvmord.

Scott, Andrea K. (2023) Refik Anadol: Unsupervised. *The New Yorker*, https://www.newyorker.com/goings-on-about-town/art/refik-anadol-unsupervised.

Seeliger, Martin and Sevignani, Sebastian (2022) A new structural transformation of the public sphere? An introduction. *Theory, Culture & Society*, 39(4): 3–16.

Seitz-Wald, Alex and Yurcaba, Jo (2023) Trump vows to 'stop' gender-affirming care for minors if re-elected for president. *NBC News*, https://www.nbcnews.com/politics/2024-election/trump-vows-stop-gender-affirming-care-minors-re-elected-president-rcna68461.

Select Committee to Investigate the January 6th Attack on the United States Capitol (2022) *Final Report.* Washington DC: US Government Publishing Office.

Serazio, Michael (2017) Branding politics: Emotions, authenticity, and the marketing culture of American political communication. *Journal of Consumer Culture*, 17(2): 225–241.

Shifman, Limor (2014) *Memes in Digital Culture.* Cambridge, MA: MIT Press.

Shklovski, Irina and Grönvall, Erik (2020) CreepyLeaks: Participatory speculation through demos. *NordiCHI '20: Proceedings of the 11th Nordic Conference on Human-Computer Interaction: Shaping Experiences, Shaping Society*, https://doi.org/10.1145/3419249.3420168.

Shome, Raka (1996) Postcolonial interventions in the rhetorical canon: An 'other' view. *Communication Theory*, 6(1): 40–59.

Smith, Jonas Heide and Just, Sine N. (2009) Playful persuasion: The rhetorical potential of advergames. *Nordicom Review*, 30(2): 53–68.

Solnit, Rebecca (2022) Greta Thunberg ends year with one of the greatest tweets in history. *The Guardian*, https://www.theguardian.com/commentisfree/2022/dec/31/greta-thunberg-andrew-tate-tweet.

Soros, George (2023) Can democracy survive the polycrisis? *Project Syndicate*, https://www.project-syndicate.org/commentary/can-democracy-survive-polycrisis-artificial-intelligence-climate-change-ukraine-war-by-george-soros-2023-06.

Sotirakopoulos, Nikos (2021) *Identity Politics and Tribalism: The New Culture Wars.* Exeter: Imprint Academic.

Sparrow, Jeff (2022) If you don't like climate activists staging art gallery protests, organise something better. *The Guardian*, https://www.theguardian.com/commentisfree/2022/oct/18/if-you-dont-like-climate-activists-staging-art-gallery-protests-organise-something-better.

Stark, David and Pais, Ivana (2020) Algorithmic management in the platform economy. *Sociologica*, 14(3): 47–72.

Stark, Luke and Crawford, Kate (2019) The work of art in the age of artificial intelligence: What artists can teach us about the ethics of data protection. *Surveillance & Society*, 17(3/4): 442–455.

Stetka, Bret (2021) An IBM AI debates humans – But it's not yet the Deep Blue of oratory. *Scientific American*, https://www.scientificamerican.com/article/an-ibm-ai-debates-humans-but-its-not-yet-the-deep-blue-of-oratory/.

Stewart, Evan and Hartmann, Douglas (2020) The new structural transformation of the public sphere. *Sociological Theory*, 38(2): 170–191.

Stiglitz, Joseph E. (2023) Inequality and democracy. *Project Syndicate*, https://www.project-syndicate.org/commentary/inequality-source-of-lost-confidence-in-liberal-democracy-by-joseph-e-stiglitz-2023-08?barrier=accesspaylog.

Storlie, Brandon (2021) The summer of Harambe: The curious case of a deceased gorilla and an animal rights campaign turned online prank. *First Monday*, 26(3), https://doi.org/10.5210/fm.v26i3.9449.

Stroud, Scott R. (2021) Connective democracy: The task before us. *Media Ethics*, 32(2): np.

Sundén, Jenny and Paasonen, Susanna (2018) Shameless hags and tolerance whores: Feminist resistance and the affective circuits of online hate. *Feminist Media Studies*, 18(4): 643–656.

Sundén, Jenny and Paasonen, Susanna (2020) *Who's Laughing Now? Feminist Tactics in Social Media*. Cambridge, MA: MIT Press.

Søgaard, Anders, Lehmann, Sune, Adler-Nissen, Rebecca, Winther, Ole and Petersen, Michael Bang (2023) Fasten your seatbelt: Intimacy capitalism is coming, https://socialsciences.ku.dk/news/2023/fasten-your-seatbelt-intimacy-capitalism-is-coming/.

Taub, Amanda (2023) Does information affect our beliefs? *New York Times*, https://www.nytimes.com/2023/08/09/world/europe/interpreter-social-media.html.

Terren, Ludovic and Berge-Bravo, Rosa (2021) Echo chambers on social media: A systematic review of the literature. *Review of Communication Research*, 9: 99–118.

Tesch, Noah (2023) Bruno Latour. *Britannica*, https://www.britannica.com/biography/Bruno-Latour.

Theiss-Morse, Elizabeth, Barton, Dona-Gene and Wagner, Michael W. (2015) Political trust in polarized times. In Brian H. Bornstein and Alan J. Tomins (eds) *Motivating Coopration and Compliance with Authority*. Cham: Springer, pp 167–190.

Thomsen, Mikkeline S.S., Steinitz, Sarah, Sivertsen, Morten Fischer and Just, Sine N. (forthcoming) Digital community centers of the 21st century? A mixed-methods study of Facebook groups as fora for connective democracy. *Social Media + Society*.

Thomson, Irene Tavis (2010) *Culture Wars and Enduring American Dilemmas*. Ann Arbor: University of Michigan Press.

Thucydides (2013) *The War of the Peloponnesians and the Athenians*. Cambridge: Cambridge University Press.

Thunberg, Greta (2019) 'Our house is on fire': Greta Thunberg, 16, urges leaders to act on climate. *The Guardian*, https://www.theguardian.com/environment/2019/jan/25/our-house-is-on-fire-greta-thunberg16-urges-leaders-to-act-on-climate.

Thunberg, Greta and Fridays for Future (2023) We won't stop speaking out about Gaza's suffering: There is no climate justice without human rights. *The Guardian*, https://www.theguardian.com/commentisfree/2023/dec/05/gaza-climate-justice-human-rights-greta-thunberg.

Törnberg, Petter, Anderson, Claes, Lindgren, Kristian and Banisch, Sven (2021) Modeling the emergence of affective polarization in the social media society. *PLOS One*, 16(10): e0258259.

Treem, Jeffrey W. and Leonardi, Paul M. (2013) Social media use in organizations: Exploring the affordances of visibility, editability, persistence, and association. *Annals of the International Communication Association*, 36(1): 143–189.

Triggs, Anthony Henry, Møller, Kristian and Neumayer, Christina (2021) Context collapse and anonymity among queer Reddit users. *New Media & Society*, 23(1): 5–21.

UNDP (2021) World's largest survey of public opinion on climate change: A majority of people calling for wide-ranging action, https://www.undp.org/press-releases/worlds-largest-survey-public-opinion-climate-change-majority-people-call-wide-ranging-action.

Van der Zwan, Natascha (2014) Making sense of financialization. *Socio-Economic Review*, 12(1): 99–129.

Van Dijck, José, Poell, Thomas and de Waal, Martijn (2018) *The Platform Society*. Oxford: Oxford University Press.

Vega, Nicolas (2023) Mark Zuckerberg calls off Elon Musk cage match: 'If Elon ever gets serious … he knows where to reach me'. *CNBC*, https://www.cnbc.com/2023/08/14/mark-zuckerberg-calls-off-elon-musk-cage-match.html.

Venturini, Tommaso (2010) Diving in magma: How to explore controversies with actor-network theory. *Public Understanding of Science*, 19(3): 258–273.

Venturini, Tommaso and Munk, Anders Kristian (2021) *Controversy Mapping: A Field Guide*. Cambridge: Polity Press.

Verma, Pranshu (2023) They fell in love with AI bots. A software update broke their hearts. *Washington Post*, https://www.washingtonpost.com/technology/2023/03/30/replika-ai-chatbot-update/.

Vickers, Brian (1989) *In Defence of Rhetoric*. Oxford: Clarendon Press.

Villadsen, Lisa S. (2020) Progress, but slow going: Public arguments in the forging of collective norms. *Argumentation*, 34: 325–337.

Villarreal, Alexandra (2021) 'Medium is the message': AOC defends 'tax the rich' dress worn to Met Gala. *The Guardian*, https://www.theguardian.com/us-news/2021/sep/14/aoc-defends-tax-the-rich-dress-met-gala.

Vincent, James (2016) Twitter taught Microsoft's chatbot to be a racist asshole in less than a day. *The Verge*, https://www.theverge.com/2016/3/24/11297050/tay-microsoft-chatbot-racist.

Vo, Lam Thuy (2020) How the internet created multiple publics. *Georgetown Law Technology Review*, 4(2): 399–412.

Von Hein, Shabnam (2022) Iranians use social media to keep protest movement alive. *DW*, https://www.dw.com/en/iranians-use-social-media-to-keep-protest-movement-alive/a-63767075.

Waqas, Ahmed, Salminen, Joni, Jung, Soon-gyo, Almerekhi, Hind and Jansen, Bernard J. (2019) Mapping online hate: A scientometric analysis on research trends and hotspots in research on online hate. *PLOS One*, https://doi.org/10.1371/journal.pone.0222194.

Warner, Michael (2002) *Publics and Counterpublics*. New York: Zone Books.

Watts, Jonathan (2021) 'Case closed': 99.9% of scientists agree climate emergency caused by humans. *The Guardian*, https://www.theguardian.com/environment/2021/oct/19/case-closed-999-of-scientists-agree-climate-emergency-caused-by-humans.

Weil, Elizabeth (2023) You are not a parrot. *New York Magazine*, https://nymag.com/intelligencer/article/ai-artificial-intelligence-chatbots-emily-m-bender.html.

Wertheimer, Tiffany (2022) Blake Lemoine: Google fires engineer who said AI tech has feelings. *BBC*, https://www.bbc.com/news/technology-62275326.

Wills, Garry (2011) The words that remade America. *The Atlantic*, https://www.theatlantic.com/magazine/archive/2012/02/the-words-that-remade-america/308801/.

Whiteman, Hilary (2023) As a child, she stitched clothes in a sweatshop. Now she sees modern slavery everywhere. *CNN*, https://edition.cnn.com/2023/05/23/asia/modern-slavery-index-2023-intl-hnk/index.html.

Willis-Chun, Cynthia (2008) No recipe for success: Avoiding critical cookie cutters. *Review of Communication*, 8(1): 29–31.

Wøldike, Morten Emmerik (2022) Had har konsekvenser: LGBT+ personer er utrygge på gaden. *The Danish Institute for Human Rights*, https://menneskeret.dk/nyheder/had-konsekvenser-lgbt-personer-utrygge-paa-gaden.

Young, Cathy (2019) (Almost) everything you know about GamerGate is wrong. *Medium*, https://medium.com/arc-digital/almost-everything-you-know-about-gamergate-is-wrong-c4a50a3515fb.

Ytre-Arne, Brita and Moe, Hallvard (2021) Folk theories of algorithms: Understanding digital irritation. *Media, Culture and Society*, 43(5): 807–824.

Zhang, Xinzhi, Lin, Wan-Ying and Dutton, William H. (2022) The political consequences of online disagreement: The filtering of communication networks in a polarized political context. *Social Media + Society*, https://doi.org/10.1177/20563051221114391.

Žižek, Slavoj (2023) The axis of denial. *Project Syndicate*, https://www.project-syndicate.org/commentary/left-right-populist-alliance-against-ukraine-by-slavoj-zizek-2023-06.

Zuboff, Shoshana (2015) Big other: Surveillance capitalism and the prospects of an information civilization. *Journal of Information Technology*, 30: 75–89.

Zuboff, Shoshana (2019) *The Age of Surveillance Capitalism: The Fight for a Human Future at the New Frontier of Power*. New York: Public Affairs.

Zuboff, Shoshana (2022) Surveillance capitalism or democracy? The death match of institutional orders and the politics of knowledge in our information civilization. *Organization Theory*, 3(3): 1–79.
Zulli, Diana, Liu, Miao and Gehl, Robert (2020) Rethinking the 'social' in 'social media': Insights into topology, abstraction, and scale on the Mastodon social network. *New Media & Society*, 22(7): 1188–1205.
Zyskind, Harold (1950) A rhetorical analysis of the Gettysburg Address. *Journal of General Education*, 4(3): 202–212.

Index

D

E

O

P

Q

R

S